# Creation

# of

# Species

# Creation

# of

# Species

## The Role of Angels
## in Intelligent Design

Jason North

Published by Jason North.
jasonnorth@jasonnorthbooks.com

Bible quotes are from the New King James Version unless
otherwise indicated. Scripture taken from the New King James
Version®. Copyright © 1982 by Thomas Nelson. Used by
permission. All rights reserved.

ISBN-13 : 979-8762474238

See author's website for announcements about future books:
http://www.jasonnorthbooks.com.

# TABLE OF CONTENTS

# CHAPTER ONE

# The Conflict

## A Cultural War

The United States is in a cultural war.

This nation was founded largely by people with deep religious beliefs. It was founded on traditional Christian values. Many who came to this land came from Europe to escape religious persecution and to seek freedom to worship God as they chose.

In the early years of this nation, many Americans tried to live moral lives. They believed in God, and they believed that the Bible was God's word. As imperfect as they were, many of them tried to live by the moral standards of the Bible as they understood them. They respected the Bible as the word of God, and they believed that the Bible had authority over their lives.

Many of the great universities of this country were also founded on religious values.

The very Declaration of Independence, which is the legal document that establishes the United States as an independent nation, acknowledges God's existence as creator of mankind when it says, "We hold these truths to be self-evident, that all men are created equal, that they are endowed by their Creator with certain unalienable Rights . . . ." I will say more about this later.

But in recent decades, many Americans have been turning away from faith in God and the Bible. Many of the universities founded on religious values have become totally secular, atheistic, having no regard for God and the Bible—even scorning and ridiculing God and the Bible.

Faith in science is replacing faith in God. Secular bias against God is diminishing Christian influence in the United States.

In 1859, Charles Darwin published his book, *On the Origin of Species by Means of Natural Selection,* popularly known today as, *Origin of Species.* He was not the first to propose that the great varieties of life evolved, but he proposed a mechanism to explain how this could have occurred. He proposed that random variation would occur from one generation to the next in species and that natural selection—the survival of the fittest— would favor those changes that made a species more fit to survive and reproduce in its environment. In this way, species could change over time. New, beneficial traits could develop and be passed on to future generations, and from these changes, new species would develop. He called this process, "descent with modification."

Over time, Darwin's book had an enormous impact. It helped to drive a trend in the scientific community away from belief in God and toward atheism. Or, to put it another way, Darwin's theory was used as a tool by those in the scientific community who wanted to promote atheism.

Today, the scientific community is entirely atheistic in its approach to understanding the world.

Not all scientists are atheists. But the scientific community, in its investigations, research, and theories, behaves as if there is no God who made the universe and can intervene in physical processes.

Scientists, in their work, follow what they call the scientific method.

The scientific method is a term scientists use for a way of doing scientific research that most scientists agree with. There are a number of principles in it that scientists are expected to follow in their professional work. This method can be articulated and described in various ways. Not all scientists necessarily adhere strictly to any method in their research and thinking. But most profess to follow the scientific method.

I won't attempt to describe all aspects of the scientific method here, but I will point out one feature that is practiced by most scientists in their work, whether it is articulated in any description of the scientific method or not.

Whether explicitly stated or not, in actual practice, one of the principles of the scientific method is that no supernatural cause is to be considered as an explanation for physical evidence scientists observe in their research and experimentation. Only natural, physical causes are to be considered as possible explanations.

Thus, scientists in their work are not allowed by the scientific method to consider that God may have designed and created the wide variety of species of life. They cannot even consider it a possibility, no matter what the evidence is. God and the supernatural realm are "off limits"—not to be considered—as far as science is concerned.

The theory of evolution and Darwin's explanation of random variation and natural selection have become whole-heartedly accepted by the scientific community as truth. Not only does science believe in evolution as dogma, it pushes acceptance of evolution very hard upon the general public. Science tries to build a culture that implies you are foolish or ignorant if you don't believe in evolution. Evolutionists will often say something like, "No well-informed person doubts evolution."

Militant atheism tries to silence or marginalize those who speak against evolution. There can be persecution against scientists who do not agree with the theory of evolution. Some who do not agree with evolution keep silent or outwardly profess agreement to protect their careers.

The alternative to evolution is intelligent design. Intelligent design is a general term for the proposition that life and the vast variety of species are not the result of random forces but have been deliberately *designed* and brought into being by an intelligent designer (or designers). That designer is understood by many to be God, but the intelligent design movement does not specify God as the designer but leaves that question open, focusing instead on examination of evidence it believes points to the *impossibility* of life resulting from random forces only, thus requiring design by one or more intelligent entities.

The vast majority of scientists, science teachers, and science writers accept evolution and reject intelligent design, at least as far as their professional work is concerned.

But that puts the science community and evolution into conflict with many people's religious beliefs. Many traditional Christian families rear their children to believe in God and the Bible. The Bible teaches that God made all things, including all the species of plants and animals. When these families send their children to public school, they are taught by their science teachers that their family religious traditions and beliefs are wrong. They say that life came about from random mutation and natural selection—evolution—not from God's creative acts as the Bible teaches.

But the Constitution and our national traditions protect religious freedom from government interference, or are supposed to, and that includes parents' freedom to teach creation to their children without government interference.

But this isn't just about children. Adult students who believe in creation because of their religious beliefs can feel pressured to express agreement with evolution in the public college classroom—in their test answers and in their homework assignments.

They can be faced with a choice to compromise with their faith to get good grades or tell the truth and risk failure. Ultimately, some students who refuse to compromise can find certain career paths in science blocked to them.

Science rejects the Bible.

And while some students feel pressured by their biology teachers to compromise with their religious beliefs, many other students are not sure, and some who enter the public school classroom believing the Bible end up rejecting it after they study the evidence their teachers claim support evolution.

Here in the United States, there is enormous interest by the general public in the creation vs. evolution controversy. There are a great number of books published on the subject. Next time you are on Amazon.com searching for some book, try entering the search terms "creation" or "evolution" and see how many books are listed.

## What Is the Truth?

What is the truth?

Science, in its interpretation of the evidence it has found, rejects the Genesis account of creation in the Bible. And there certainly is a conflict between science's explanation for life in all its variety and species and the Genesis account. But is there really a conflict between the actual physical evidence science has discovered and the Bible?

Can the Bible be reconciled with the physical evidence science has discovered about the origins of biological life in all its variety?

The theory of evolution is more than a collection of evidence. It is the scientific community's *interpretation* of that evidence that makes the theory of evolution what it is. That interpretation cannot be reconciled with the Bible. But what about just the evidence itself? If properly interpreted and understood, can the evidence be reconciled with the Genesis account of creation?

In this book, I am going to review that evidence as well as the account of creation in the Bible and show how the physical evidence of fossils, chemistry, and genetics that science uses to support its theory of evolution can be reconciled with a *literal* understanding of the creation account in Genesis.

Most people have not heard this explanation before. Do not expect it to be something you are familiar with. You may be surprised.

The truth is, both science and most religion have made mistakes in understanding the origin of the species. Many churches and religious groups have misunderstood Genesis, and science has misinterpreted the physical evidence it has discovered.

In this book, I am going to closely examine the Bible account of creation and the physical evidence science has discovered and show what is probably the *one* way—the only way—they can be reconciled.

# CHAPTER TWO

# What Do the Bible and Genesis Really Say?

## How Should We View the Genesis Account?

What does Genesis really say about creation?

Most traditional religion has not understood this, and that is part of the problem.

Genesis Chapters 1 and 2 give the account of the six days of creation.

There are a number of interpretations of this account held by those who profess belief in the Bible.

One interpretation is that the six days are symbolic, not literal 24-hour days, and that the entire account is symbolic. According to this interpretation, the six days could represent hundreds of millions of years. People who hold to this interpretation say that the Bible is true and the Genesis account is accurate, but God could have used, or allowed, evolution as His way of creating the species of life. Thus, there is no conflict between science and the Bible because the Bible can be interpreted symbolically to represent anything that science believes about the history of life and the origin of species.

There are varieties of views within this overall interpretation, and not all views accept everything that
science teaches, but what they have in common is the
view that the first chapters of Genesis should not be tak-
en literally.

# The Bible Claims Authority over What We Believe

Should the Bible be taken literally or symbolically?

Some people say, symbolically. They say that the account of the six days of creation is simply a metaphor
that teaches us that God is overall creator of the earth
and the universe.

The reason many people take this view is that they
know of no other way to reconcile the Bible with the evidence science has discovered.

But there is a problem with that.

There are many passages in the Bible that teach that
the Bible is God's word and is infallibly correct. The
Bible teaches that all scripture is inspired by God and is
God's word—God speaking. It further teaches that God
cannot lie and scripture cannot be broken. And it
teaches the importance for Christians of trusting God
and believing what He says (John 10:35; Titus 1:2;
Hebrews 6:18; Proverbs 3:5; Psalm 19:7; Hebrews 3:17).

That does not mean, by itself, that a symbolic interpretation of the Bible is wrong. But in order to examine the issue of taking the Bible literally or symbolically, we must first establish the principle that we need to respect and believe God's word. "But on this one will I look: On him who is poor and of a contrite spirit, and who trembles at My word" (Isaiah 66:2). Simply saying, "The Bible is wrong," is not an option for Christians.

"All Scripture is given by inspiration of God, and is profitable for doctrine, for reproof, for correction, for instruction in righteousness" (2 Timothy 3:16).

"For what does the Scripture say? 'Abraham believed God, and it was accounted to him for righteousness' " (Romans 4:3).

"The entirety of Your word is truth" (Psalm 119:160). "Sanctify them by Your truth. Your word is truth" (John 17:17).

Moreover, the Bible is clear that God *requires* that we have the same attitude of belief in God and His word that Abraham had.

"Listen to Me, you who follow after righteousness, You who seek the LORD: Look to the rock from which you were hewn, And to the hole of the pit from which you were dug. Look to Abraham your father, And to Sarah who bore you; For I called him alone, And blessed him and increased him" (Isaiah 51:1–2).

"Therefore He who supplies the Spirit to you and works miracles among you, does He do it by the works of the law, or by the hearing of faith? just as Abraham 'believed God, and it was accounted to him for righteousness.' Therefore know that only those who are of faith are sons of Abraham. And the Scripture, foreseeing that God would justify the Gentiles by faith, preached the gospel to Abraham beforehand, saying, 'In you all the nations shall be blessed.' So then those who are of faith are blessed with believing Abraham" (Galatians 3:5–9).

It is vitally important for Christians that we respect and believe what God says in His word, the Bible. Therefore, we cannot lightly take to be symbolic or figurative what God says unless we have good grounds for doing so. If the language is literal and a normal reading of it would be understood as literal, then we must believe what it literally says.

## The Proof of the Bible

How can one know that the Bible is God's word?

Anyone can choose to believe that the Bible is false. Most who believe this believe that there is no God who inspired the Bible and that the Bible is no more than a collection of writings by uninspired human authors and full of mistakes. There may also be some who believe the Bible is God's inspired word but that God does not always tell the truth, though this view is probably less common.

One can make a choice to trust God, that He will always tell the truth, or not, but to make that choice regarding the Bible, one must first know that the Bible is inspired by God—that it is God's word—that it is God speaking.

Can the Bible be proved to be the inspired word of God?

Yes.

There are two basic proofs. One is the internal consistency of the Bible. The other is fulfilled prophecy.

An atheistic skeptic may laugh at the first proof—internal consistency. If a skeptic has read the Bible, or critical reviews of it, he is likely to think he sees many inconsistencies and contradictions. He looks for them. And he is glad when he finds what he thinks are contradictions. That is his attitude. He doesn't consider ways that different passages in the Bible can be reconciled and may compliment each other. He is not looking for explanations. He wants to find contradictions, and what he looks for, he thinks he finds.

But the Bible is consistent when properly understood.

Not only that, but the Bible is consistent in surprising ways, ways that cannot be accounted for if the Bible is just a collection of writings of independent authors who were not inspired by God. There are passages in the Old Testament, for example, that seem to make no sense apart from the New Testament—but then they become clear. There are events and teachings in the Old Testament whose reasons are mysterious or unclear—yet reasons for them are made clear in the New Testament.

Only the inspiration of God can account for the consistency of the Bible as a whole.

Then there is also the proof of fulfilled prophecy.

Prophecies have been fulfilled in the last couple of hundred years that were written thousands of years ago—fulfilled in a way that cannot be plausibly explained as coincidence to a reasonable mind.

One example is a passage in Daniel about the times we are living in right now.

There is evidence in the Bible for a seven-thousand year plan of God for the salvation of mankind. The weekly cycle of six work days followed by a Sabbath day of rest is God's model of that plan, with each day representing a thousand years. "But, beloved, do not forget this one thing, that with the Lord one day is as a thousand years, and a thousand years as one day" (2 Peter 3:8).

When Adam sinned and followed Satan, he set the course for the world that followed (Genesis 3:1–19). Satan has ruled this world since that time, and God has allowed Satan to rule the earth for a purpose, which is to teach mankind the lesson that Satan's way produces suffering, destruction, and death. Satan can only do what God allows him to do. Nevertheless, God allows Satan to blind the majority of the world to God's truth during the six-thousand year period since Adam's sin. I will talk about that in more detail later in this book.

But at the end of the six thousand years, Christ will return to this earth as King of kings and Lord of lords to rule the earth for one thousand years (Revelation 20:1–6). Satan will be bound during that time and not allowed to deceive and tempt the human race into sin (Revelation 20:1–3). The true gospel will be preached all over the world without Satan around to deceive and blind mankind into rejecting or misunderstanding the gospel. And the world that will result will be a world of unprecedented peace, happiness, and joy all over the earth (Micah 4:1–4; Amos 9:13–15; Isaiah 9:6–7; 11:1–5; 25:6–8; 35:5–7; Psalm 98:4–9).

But here is what I want you to notice.

We are near the end of the six-thousand year age of man. The chronology of the Bible can be calculated by the records of the ages of the men who lived and how old they were when their sons were born and by other Bible passages that help establish Bible dates. This can be done to date events up to the time when biblical events can be matched with events from secular history, which dates we know. This calculation is approximate because ages of men are rounded to the nearest year. Nevertheless, we can know that it has been approximately six thousand years since God made Adam. James Ussher based his date of the creation of Adam in 4004 BC on this type of calculation, though we must regard this as approximate.

So we are near the end of six thousand years of Satan's rule of the earth—the end of the age of man under Satan's influence.

Now look at a prophecy in Daniel about what would happen in the "time of the end," a term used to refer to the time near the end of this age.

Notice what Daniel wrote, under inspiration of God, in Daniel 12:4: "But you, Daniel, shut up the words, and seal the book until the time of the end; many shall run to and fro, and knowledge shall increase."

Near the end of the six thousand years, knowledge and transportation (people running "to and fro") would increase.

Has that happened?

In the last two hundred years, and even more so in the last one hundred years, there has been a knowledge explosion unlike anything seen in human history before. That knowledge explosion has resulted in a vast increase in transportation. People routinely fly overnight between the United States and Europe, making a journey in a night that would have taken months only a few hundred years ago.

If Daniel was not inspired by God, how could he have predicted that at the end of an age as indicated by Bible chronology there would be such a knowledge explosion as we have witnessed in our lifetimes? What are the chances that this could be a coincidence—that this explosion in knowledge would come just near the end of six thousand years of man's history, and not five hundred years ago, or a thousand years ago, or a thousand years from now in the future? The chances of that would be slim. But if God inspired Daniel's prophecy and also the recording of Bible chronology that gives about six thousand years as the age of man's society, then this convergence of the actual knowledge explosion we have witnessed with Bible prophecy is expected, and this serves as evidence that God inspired the Bible and that the Bible is God's word—it is God speaking.

There are other prophecies that prove the inspiration of the Bible, prophecies given thousands of years ago that have been accurately fulfilled in just the last two hundred years—prophecies that prove that the Bible is the word of the Creator, for no man could predict these events thousands of years in advance. The prophecy in Daniel is just one example.

With that background, that we can prove that the Bible is inspired by God and is God speaking, and that we need to have profound respect for and belief in what God says, let's now examine if the creation account in Genesis should be understood literally or figuratively.

## Are the Six Days of Creation Literal 24-Hour Days?

How are we to understand the six days of creation?

Could God be speaking symbolically or figuratively?

How does one honestly communicate?

If we choose to believe God's word, then we know that God cannot lie to us. If He says something in the Bible, we can know that it is true.

And speaking symbolically or figuratively is not necessarily lying. One can use a metaphor to tell the truth. We do it all the time in our conversations with each other.

A person who has a congested or scratchy throat may say, "I have a frog in my throat." If it is raining heavily, one might say, "It is raining cats and dogs." Those are not lies, because we understand each other. We know when the other person is using figurative language.

And that is the key.

We are not lying when we speak symbolically if we know the other person understands what we really mean. One might say to a friend about someone he is angry with, "I'm going to kill him!" But he knows his friend understands he is not speaking literally. And if he sees by a shocked expression on his friend's face that he was taken literally, he clarifies: "I don't mean that literally."

God will not use metaphors and figures of speech to trick us or deceive human beings. He won't use symbolic language to lie to us, knowing people would take him literally. When He uses symbols in the Bible, He makes sure we know they are symbols, and He usually lets us know what those symbols represent.

If God used language that could be taken either literally or figuratively, and we could not tell which is intended, we could never be sure what He is telling us.

Here is an example of God's use of symbolic language.

After God brought Israel out of Egypt, He said to them, "You have seen what I did to the Egyptians, and how I bore you on eagles' wings and brought you to Myself" (Exodus 19:4).

Now, God was using a metaphor. He was not speaking literally when He said He took Israel out of Egypt on eagles' wings. We know that because we can read the account of *how* God took Israel out of Egypt. He made it possible for them to walk out of Egypt.

And Israel knew it was a metaphor the same way.

So there was no chance of a misunderstanding. God was not lying. He knew He would not be misunderstood. He knew that both ancient Israel and we as we read the Bible would understand that eagles' wings were a metaphor, not a phrase to be taken literally.

But that is not the case in the account of the six days of creation.

There is nothing in the first chapter of Genesis to suggest that the six days were a metaphor, not to be taken literally. Neither is there anything elsewhere in the Bible to suggest the possibility of symbolic language in the creation account. This account would no doubt be taken literally by anyone who read the Bible in ancient times and many who read the Bible today.

God knows that people would take this account literally, and He would not lie to them. He won't lie, then hide behind the excuse, "I was speaking metaphorically."

It is only because of recent scientific evidence that some doubt the literal accuracy of Genesis chapter one. They know of no way to reconcile the evidence with what this passage actually says. So they try to maintain some level of belief in the Bible by saying, "It is a metaphor. It could represent millions of years."

But that is wrong on two accounts.

First, as I said, God will not use figurative language if He knows that men will take what He says literally and be deceived. He won't use metaphors to lie to us.

But second, there *is* a way to reconcile a literal interpretation of Genesis chapter one with the evidence science has found. We don't have to resort to symbols.

How to reconcile scientific evidence with the teaching of the Bible about creation is the subject of the rest of this book.

## What Does Genesis Chapter One Really Say?

Let's not overlook the fact that the six days of creation start with the existence of the planet earth.

"In the beginning God created the heavens and the earth. The earth was without form, and void; and darkness was on the face of the deep. And the Spirit of God was hovering over the face of the waters" (Genesis 1:1–2).

God made the physical planet earth *before* the six days. It was already in existence, in darkness and covered with water. Did God originally make it that way?

The words "without form and void" are translated from the Hebrew words tohu and bohu. They mean to lie waste in confusion and emptiness.

But God is not the author of confusion. "For God is not the author of confusion but of peace, as in all the churches of the saints" (1 Corinthians 14:33). Why would God originally make the earth in such a state? Or did He?

There is a verse in Isaiah that will help answer our question.

"For thus says the LORD, Who created the heavens, Who is God, Who formed the earth and made it, Who has established it, Who did not create it in vain, Who formed it to be inhabited: 'I am the LORD, and there is no other'" (Isaiah 45:18).

That phrase "in vain" is translated from the same Hebrew word, "tohu," that is translated "without form" in Genesis 1, verse 2. God did not create the earth in a state of "tohu."

Genesis 1:1 says that God created the earth. Isaiah 45:18 says God did not create the earth in a state of "tohu." Genesis 1:2 says the earth was in a state of "tohu." Evidently, the earth *became* "tohu," in a state of waste and confusion, in a state of desolation, in darkness and covered with water, between verses 1 and 2 of Genesis chapter 1.

Also, that word "was" in this context can have the meaning of "became."

So Genesis does not say that God originally made the earth "without form and void" as described in verse 2.

God did not originally create the earth in a state of desolation as described in Genesis 1:2. But something happened to produce that state after God originally made the earth.

Genesis is the story of creation and the beginning of human history, but it is only a summary account. It doesn't give every detail of what happened. But we can infer some things from other passages in the Bible.

God put the Bible together to be read and studied as a whole. Details of lessons in one part of the Bible are sometimes explained in another part of the Bible.

What could cause the earth to come to be in a desolate state as described in Genesis 1:2?

There is an example of an event that destroyed the earth, also described in the book of Genesis. I am referring to the flood in Noah's day.


## The Example of the Flood in Noah's Day


After God made Adam and Eve, the human race began to multiply. But men were evil, and there was much violence and sin. "Then the LORD saw that the wickedness of man was great in the earth, and that every intent of the thoughts of his heart was only evil continually. And the LORD was sorry that He had made man on the earth, and He was grieved in His heart. So the LORD said, 'I will destroy man whom I have created from the face of the earth, both man and beast, creeping thing and birds of the air, for I am sorry that I have made them'" (Genesis 6:5–7).

Because of worldwide sin, God decided to destroy the surface of the earth and all air-breathing life with a flood, saving only Noah and his family and pairs of the various air-breathing land animals to populate the earth afterwards.

"And God said to Noah, 'The end of all flesh has come before Me, for the earth is filled with violence through them; and behold, I will destroy them with the earth'" (Genesis 6:13).

"And behold, I Myself am bringing floodwaters on the earth, to destroy from under heaven all flesh in which is the breath of life; everything that is on the earth shall die" (Genesis 6:17).

"Now the flood was on the earth forty days. The waters increased and lifted up the ark, and it rose high above the earth. The waters prevailed and greatly increased on the earth, and the ark moved about on the surface of the waters. And the waters prevailed exceedingly on the earth, and all the high hills under the whole heaven were covered. The waters prevailed fifteen cubits upward, and the mountains were covered. And all flesh died that moved on the earth: birds and cattle and beasts and every creeping thing that creeps on the earth, and every man. All in whose nostrils was the breath of the spirit of life, all that was on the dry land, died. So He destroyed all living things which were on the face of the ground: both man and cattle, creeping thing and bird of the air. They were destroyed from the earth. Only Noah and those who were with him in the ark remained alive" (Genesis 7:17–23).

Here we see a case where God destroyed life on earth by bringing destruction to the surface of the earth. Why? Because of *sin*. Sin can be a cause of destruction. We could add the examples of the destruction of Sodom and Gomorrah (Genesis 19:20–21, 24–25), the warning to Nineveh (Jonah 3:4–10), the promises of blessings for obedience and curses for disobedience given to Israel in the book of Deuteronomy (Deuteronomy 7:12–26; 8:1–20; 28:1–68), and the Day of the Lord when God will punish the whole world for its sins (Joel 1:15; Isaiah 13:9–13).

Could sin have been the cause of an earlier destruction that occurred between verses 1 and 2 of Genesis chapter 1? Could the earth have become "without form and void," in darkness and covered with water, as a result of sin?

If so, it would not have been the sin of man. Man did not exist at that time.

But before God made man, He made angels.

## Lucifer and the Angels on the Earth

There is a Bible passage that shows that angels existed before the earth existed.

"Where were you when I laid the foundations of the earth? Tell Me, if you have understanding. Who determined its measurements? Surely you know! Or who stretched the line upon it? To what were its foundations fastened? Or who laid its cornerstone, When the morning stars sang together, And all the sons of God shouted for joy?" (Job 38:4–7). "Morning stars" and "sons of God" refer to angels.

Why did the angels sing and shout for joy at the creation of the earth?

The earth must have been beautiful in its way, even before biological life was made. There would have been beautiful mountains and rock formations, blue sky with white clouds, blue streams, lakes, and oceans, and beautiful sunrises and sunsets. The earth was not in darkness and chaotic.

And the earth was to be the home of Lucifer and many angels.

There are two passages about Lucifer, who became Satan the Devil, in the Old Testament that shed some light on his history and character. And there is a passage in Revelation that indicates that he led one-third of God's angels into sin, and they became demons.

One of those passages indicates that Lucifer was on the earth.

"How you are fallen from heaven, O Lucifer, son of the morning! How you are cut down to the ground, You who weakened the nations! For you have said in your heart: 'I will ascend into heaven, I will exalt my throne above the stars of God; I will also sit on the mount of the congregation On the farthest sides of the north; I will ascend above the heights of the clouds, I will be like the Most High'" (Isaiah 14:12–14).

Notice that Lucifer was on the earth, for verses 13 and 14 quote him as saying, "I will ascend into heaven" and "I will ascend above the heights of the clouds."

The other Old Testament passage that sheds light on Lucifer's character is in Ezekiel chapter 28: "Son of man, take up a lamentation for the king of Tyre, and say to him, 'Thus says the Lord GOD: "You were the seal of perfection, Full of wisdom and perfect in beauty. You were in Eden, the garden of God; Every precious stone was your covering: The sardius, topaz, and diamond, Beryl, onyx, and jasper, Sapphire, turquoise, and emerald with gold. The workmanship of your timbrels and pipes Was prepared for you on the day you were created. You were the anointed cherub who covers; I established you; You were on the holy mountain of God; You walked back and forth in the midst of fiery stones. You were perfect in your ways from the day you were created, Till iniquity was found in you. By the abundance of your trading You became filled with violence within, And you sinned; Therefore I cast you as a profane thing Out of the mountain of God; And I destroyed you, O covering cherub, From the midst of the fiery stones. Your heart was lifted up because of your beauty; You corrupted your wisdom for the sake of your splendor"'" (Ezekiel 28:12–17).

Lucifer was not alone in his rebellion, for Revelation indicates that he drew a third of the angels to follow him, and they became demons. "And another sign appeared in heaven: behold, a great, fiery red dragon having seven heads and ten horns, and seven diadems on his heads. His tail drew a third of the stars of heaven and threw them to the earth" (Revelation 12:3–4). The "stars of heaven" who fell to the earth are fallen angels, demons, who joined Satan in his sin, apparently a third of all angels.

It was the sin and rebellion of Lucifer, who became Satan, and his angels, who became demons, that resulted in the destruction of the surface of the earth— causing it to be in darkness and covered with water—in a state of chaos and confusion, as described in Genesis 1:2, just as the sins and violence of the human race later resulted in the destruction of the surface of the earth by the flood in Noah's day (Genesis 6:13).

How did this destruction come about? The Bible doesn't say. It could have come from God as punishment for the wickedness of Satan and the demons just as God punished mankind for their wickedness with a flood in the days of Noah. Or, it may have been Satan's act to destroy God's creation, which God allowed.

Did this destruction come immediately as soon as Lucifer began to sin? No, probably not. God did not immediately destroy mankind in a flood as soon as Adam sinned or as soon as Cain killed Abel. The wickedness and violence of mankind built up and continued for a while before God finally destroyed mankind in a flood.

Likewise, the sins of Satan and the demons may have continued for a while, perhaps for hundreds of millions of years, before the destruction of the surface of the earth occurred. And it may have taken time for Satan to corrupt one-third of the angels.

Lucifer started out righteous. God did not create him as an evil being. He started out "perfect in his ways." And he continued perfect in his ways, *until he sinned*. See Ezekiel 28:15.

Apparently, Lucifer was the first spirit being to sin. He then led the angels under his authority and supervision on the earth to sin, and when they sinned, they became demons. See Revelation 12:3–4.

It might have taken time for Satan to persuade a third of the angels to follow his rebellion and sinful ways, perhaps millions of years. And even after they sinned, it may have taken millions of years for the sin to reach the point where destruction of the surface of the earth was the result, just as it took time for the sins of mankind to result in the flood in Noah's day.

Why did God place Lucifer and some of the angels on the earth in the first place? I will explain that in a later chapter. They were given a job to do. And they were also being tested.

So the sequence seems to be this. God originally created the earth, not in chaos and confusion, but in beauty. God placed Lucifer and a portion of the angels— perhaps one-third—on the earth to do a job, a job that would give them interesting work to do—work that would occupy their minds and abilities and give them an opportunity to serve God and God's purpose, but also work that would test their obedience and loyalty to God. Lucifer was leader over the other angels (remember, he had a throne on the earth—Isaiah 14:13). Lucifer was perfect in his ways for a while—the Bible does not say how long—until sin was found in him—he obeyed for a time.

But he had free moral agency. God did not force him to remain righteous. He had to choose good or evil.

And at some point in time, he sinned. Iniquity was found in him. He chose the way of pride and vanity. His mind became evil. He rebelled against God. And he led a third of God's angels—probably most of the angels under his supervision on the earth—to join him in sin and rebellion against God. This may have taken a long time.

And it may have been a long time, even after the angels on the earth sinned and became evil, until the sin reached the point where the surface of the earth was destroyed and the condition described in Genesis 1:2 came to be.

Then—after a time (the Bible doesn't say how long)— God renewed the face of the earth as described in Genesis 1:3–31. God restored the earth to a state of beauty and prepared it for man. And in the sixth day, He made man, and that was the beginning of the human race and the history of the earth that is recorded in the rest of the Bible.

The Bible does not indicate how long the earth remained "without form, and void"—in chaos and confusion, in darkness and covered in water—before God began the six days of creation. As far as the Bible is concerned, it could have been a day, many days, or a number of years.

## The Earth Is Not Necessarily Six Thousand Years Old

Thus, the Bible does not say that the earth is only six thousand years old. As far as Genesis is concerned, it could be hundreds of millions of years old, or billions of years old.

The six days of creation described in Genesis 1:3–31 are six literal 24-hour days. They begin and end at sunset as God counts days (Leviticus 23:32). God used these six days to repair the damage to the surface of the earth, to restore plant and animal life, and to prepare the earth for man. And in the sixth day, God made man. And this restoration of the surface of the earth and the making of man occurred about six thousand years ago.

The fact that the description of these days includes, "So the evening and the morning were the first day" (Genesis 1:5), shows that these are actual, 24-hour days, not figurative periods of indefinite time.

The human race is about six thousand years old. But not the planet earth itself. The earth may be billions of years old.

To better understand this, it might be useful to consider the makeup of the planet earth from the inside and ask when this structure came about. The earth is made up of a core, mantle, and crust. The core is made up of an inner core, made of iron and nickel and is solid, and an outer core which is liquid. Above the core is mantle, which is liquid rock, which is known as lava when it comes from a volcano. On top of the mantle is the crust, a solid layer.

If God created the earth in six days, in which of those days did He create the core? In which day did He create the mantle and the crust? The answer is, those parts of the earth and the entire earth were made *before* the six days of creation in Genesis 1:3–31.

There is a sense in which God made the heavens and the earth in six days, and that is that in six days God renewed or refreshed the surface of the earth and made the heavens visible on the surface of the earth. That is what Genesis chapter 1 shows, and that is called "making" the heavens and the earth in Exodus 20:11. But is was not during the six days that the planet earth was brought into existence. In six days God renewed or refreshed the face of the earth (Psalm 104:30).

The Bible, even when understood literally, allows for a potentially vast amount of time between verses 1 and 2 of Genesis chapter 1—maybe billions of years—and much could have happened on the earth during that time. The Bible does not reveal much detail about what happened during this period, but scientific evidence may.

# CHAPTER THREE

# The Scientific Evidence

## Evolution Is Not a "Fact"

Sometimes, when you ask evolutionists if they can prove evolution occurred, they reply that science does not deal in "proofs."

Mathematics deals in proofs, and courts of law deal in proofs, as in, "a person is innocent until proven guilty," or, "proof beyond reasonable doubt." But evolutionists require that evolution be accepted without "proof."

Evolutionists love to say, dogmatically and with emotional emphasis, "evolution is a *fact*," as if their sureness and passion in the way they say it should convince us.

In scientific terminology, the word "fact" should not be used this way, even if evolution is true. The word "fact" in scientific literature refers to *direct* observations, usually of a simple nature, which, when put together, depending on how those observation are interpreted, can lead to conclusions and theories. So in evolutionary theory, a "fact" would be a particular fossil discovered in a particular place. Or it might be a particular DNA sequence, or a particular result in the laboratory, or a particular observation in nature. Polar bears being white and living in an environment dominated by snow and ice would be a "fact."

Facts are then interpreted and used to produce theories which explain and give reasons for the facts. You can also say that these observed facts are the evidence upon which theories are built.

I can use the theory of special relativity as an example to illustrate this. Special relativity is not controversial. Almost everyone who has a scientific background in physics agrees that the theory of special relativity is true, and I also agree that it is true. But even though it is almost universally accepted as true, it is still called a theory, not a "fact." Scientific literature would not generally refer to special relativity as "fact," unless it is written for the layman. A fact relating to special relativity would be the result of the Michelson-Morley experiment, which showed that the speed of light as measured is the same in all directions. Another fact might be the difference in time related by atomic clocks that have traveled at high speed in orbit compared to atomic clocks that remain stationary. The release of energy and reduction of mass when the nucleus of an atom decays would be another fact. The generation of new particles when a particle in a particle accelerator collides with a target is another fact that supports the theory of special relativity. But relativity, though true, is not called a "fact."

Yet evolutionists like to call evolution a fact, not a theory, because the word "theory" implies that it might not be true.

What evolutionists really mean when they say evolution is a fact is simply that *they* are sure it is true.

Why are they sure it is true? Why do they think you and I should accept it as truth?

Part of the reason hard-core evolutionists, many of them, believe evolution is true is because they *choose* to believe it is true. They want to believe it because they do not want to believe in a personal God who made this universe and intervenes in physical processes when He chooses. They do not want to believe in a creator God or in intelligent design. If they believed in God, there would likely be consequences for their state of mind and how they live their lives, consequences they do not want to face. I do not say this is the case with everyone who accepts evolution, but I believe it is true for many of the most passionate advocates of evolution.

But if you ask them how they know evolution is true, they will say that the physical evidence (observable facts) *points* to the conclusion that evolution is true.

Is that the case? Does the physical evidence that science has discovered—fossils, genetics, chemistry, comparative anatomy of species, etc.—really point to evolution?

## Evidence Must Be Interpreted

In examining this question, there is something to keep in mind.

Evidence does not point to anything until it is interpreted. There is no sign painted on the side of a polar bear with an arrow pointing to the word, "evolution." Someone may say that polar bears evolved white fur to blend in with the white background of snow and ice because a random mutation resulted in white fur and natural selection gave a reproductive advantage to white bears, but that conclusion requires *interpretation* of the evidence. The interpretation may be true or false, but the evidence itself is blind apart from interpretation. Somebody has to think about the evidence and explain what it means before it points to anything.

The interpretation of the evidence for origins of species by the scientific community is biased and unreliable from the start. Why? Because science is biased against belief in God or even the possibility that God exists. The majority scientific community is atheistic in their scientific work and teaching.

The scientific method, as practiced by majority science, does not allow supernatural explanations for physical evidence. Scientists in their work cannot acknowledge the spirit realm. If a scientist, or a science teacher in a public school, suggested that God caused a certain physical result, he or she would be laughed (or angrily driven) out of his or her job.

The scientific community only considers natural physical causes for physical evidence. This is a limitation on its ability to consider all possible causes for the physical evidence it observes. But the scientific community does not think it is a limitation, because the scientific community is dominated by atheists. Exclusion of explanations and interpretations of evidence involving supernatural and spiritual influences doesn't seem like a limitation to atheists, because they don't think that any supernatural or spiritual influence exists.

But they can't prove that. They choose to believe that the supernatural doesn't exist, but they can't prove it. It is a matter of faith for them. They won't call it faith, but that is what it is. I am using the word "faith" here as simply a chosen belief system. Atheists, in or out of science, cannot prove there is no God. They *choose* to believe it as a matter of faith.

And this is a faith that many of them are zealous to try to impose on others. Militant atheists want to impose their atheistic faith on students through the teaching of evolution.

But actually, excluding the possibility that God exists and has intervened to create the species of life we see is certainly a limitation. The scientific community is not considering all possible explanations for the evidence they observe.

Common sense indicates that if you are to correctly interpret evidence and reach a right conclusion, you must not be biased. Courts of law understand this. You need to consider all possibilities without prejudice. But atheists do not think God's existence is possible. And they do not want you to think His existence is possible. So they do not think they are excluding any possibilities when they only consider physical explanations for the evidence they observe. But they are.

## Parts of Evolutionary Theory

In order to know if the evidence points to evolution, we must first understand exactly what evolution is. We must know what the theory of evolution teaches.

Technically, the actual start of the first life on earth is not part of the theory of evolution—evolution is about the change of that first life form into multiple life forms—but as a practical matter, the issues are the same and they can be studied together. If natural forces could produce the first life form, perhaps they could produce the huge variety of life we see today, but if it took an intelligent designer to make the first life, then that intelligent designer was probably involved in the branching out of that life into a great variety of species.

The theory of evolution can be divided into several parts, and this can be done in different ways. For my purposes, I will point out three parts of the theory of evolution. Two of these parts may be right, perhaps with modification, but the third is not.

One part is common descent. The theory of evolution teaches that the wide variety of species of life, including human life, came through descent from a common ancestor. Thus, all life is related through a kind of family tree. Horses and zebras have a common ancestor. Humans and chimpanzees have a common ancestor. And going back further, the ancestor of horses and zebras and the ancestor of humans and chimps themselves have a common ancestor. This can be called, "descent with modification."

Species branched into different species, and this is how all species came to be.

A second part of evolution is that this overall family tree of species emerged gradually over millions of years, even hundreds of millions of years. It did not happen quickly, but gradually.

A third part of evolution is that the changes that resulted in new species occurred naturally, not caused by God or any spiritual or supernatural influence. This is, as I have shown, required by the scientific community practicing the scientific method. As that method is practiced, no supernatural explanation for evidence is allowed. So science's explanation for the emergence of new species must be entirely natural and material—due to physical causes only.

What is that explanation? Darwin led the way. Science teaches the mechanism of random mutation (or random variation) and natural selection.

When an organism reproduces, its genetics, which determine its characteristics, are passed on to its offspring. But sometimes there is a change in the genes that are passed on. There are mutations due to background radiation, DNA copying errors during reproduction, or other natural causes. When this happens, the offspring can be different from its parents, and those differences can be passed on from generation to generation from that time on.

In most cases, the mutation is harmful and the offspring is less fit to survive and reproduce. It tends to die out. But sometimes the mutation is beneficial, and when that happens, the offspring is better fit to survive and reproduce. That offspring tends to survive and spread, and that can be the beginning of a new species. Little by little, changes can occur until a new species is formed.

These are three parts of evolution I want to talk about in more detail. These need to be examined individually. If one or two are true, that does not mean all three are true.

And all three are needed for evolution. If one of the three is omitted, you do not have evolution as it is currently taught.

Some may believe in a kind of "theistic evolution." Perhaps that term may mean different things to different people. Some may apply that term to the idea that species came through natural descent (the first part of the theory I described) over millions of years (the second part of the theory), but by the intervention and design of God, not by random mutation and natural selection (the third part of the theory).

Whether that idea is true or not, *it is not evolution as taught by the scientific community today.* It is not evolution as taught to students and our children in the public schools. Some may call it a kind of "evolution," but it is a different animal altogether from the evolution the scientific community teaches.

Unless we can determine that the evidence, when interpreted without bias, strongly points to all three of these parts and does not allow any alternative to these parts, we cannot honestly say that the evidence points to evolution. And in determining this, we must look at the possible explanations without bias. If we want the truth, we must not limit ourselves, as majority science does, by refusing to look at the possibility that God exists and has intervened in physical events.

We need to find if the evidence points to common descent and no other explanation is plausible. We need to find if the evidence points to emergence of species over a long time span, millions of years, and no other explanation is plausible. We need to determine if the evidence points to natural physical causes—random mutation and natural selection—as the *only* causes of new life forms and new species over time. If we cannot determine that the evidence points to all three parts and no alternatives are plausible, we cannot honestly say the evidence points to evolution.

Let's examine each of these three parts in more detail.

## Emergence of Species through Common Descent

Evolutionists point to a number of pieces of evidence that they say indicate common descent. Some of their arguments are strong, some weak.

One is comparison of similar species. Evolutionists say that similarities often indicate relatedness and that species are similar because they come from a common ancestor. Some of these species, dogs and wolves for instance, are so similar they can cross-breed with each other. The similarities include anatomy, but also body chemistry, behavior, and any other characteristics. These similarities can be termed a "nested pattern of resemblance."

This is a weak argument, for God may simply have designed similar species. Similarity may indicate a common designer as much as descent from a common ancestor.

Another argument is the similarity of genetic nuclear DNA between similar species. Nuclear DNA is the DNA in the nucleus of every cell. Part of this DNA makes up genes, segments of DNA that specify proteins. It is the genetic DNA that primarily determines the characteristics of an organism. It is like a blueprint for the plant or animal. It is like computer code. It determines the characteristics of an organism as computer code determines the characteristics of a computer software program.

Species that are similar in body structure, chemistry, and behavior are also similar in their genes. Evolutionists will point to this similarity as evidence of common descent. The interrelatedness of species is shown by the similarity of their genes. The more closely related the species, the more similar their genes are.

But this also is a weak argument for common descent, for a design advocate will say that the genes of similar species were coded to be similar by the designer to produce similar body structure, chemistry, and behavior.

But not all DNA that is passed from one generation to the next is in the genes where it influences the characteristics of the organism. And even in the genes, there are differences in DNA coding that are not necessarily reflected in the structure, chemistry, or behavior of the organism.

There is evidence in the similarities and differences in DNA between species that show interrelatedness—common descent—that cannot be easily explained by intelligent design. These DNA differences do not necessarily affect the organism, yet are passed from one generation to another and can show how far back two species had a common ancestor. Design does not explain them because they apparently do not affect the characteristics of the organism. But random mutations and common descent do explain them.

There is a pattern of DNA similarities and differences between species indicating how distantly in time they had a common ancestor. This is the basis of what scientists term the "molecular clock."

When organisms reproduce, their DNA is reproduced and passed on to their offspring. Sometimes mutations occur—changes in the DNA that make the DNA of the offspring slightly different from that of its parents. These changes due to mutations are then passed on to the next generation, and the next, and so on. In this way, DNA gradually changes over time as mutations occur and they are passed from generation to generation. How much the DNA changes depends on how long it has been changing—the longer the time period and the more generations have reproduced, the greater the DNA differences from a distant ancestor.

Thus, the differences in DNA between two existing species today can indicate how long ago they had a common ancestor. Even if there are problems in determining in absolute years how long ago they had a common ancestor, this method can be used to compare how far back two species had a common ancestor with how far back one of those species had a common ancestor with a third species.

It is not as complicated as my convoluted sentence above may indicate (I am not the best writer on the planet).

Compare DNA of a fox with a wolf and note the differences. Foxes and wolves are believed, based on body structure, chemistry, and behavior, to be closely related. Both are canines. They are thought to both come from a common ancestor not too many millions of years ago. Now compare the DNA of one of those animals—let's say the wolf—with the DNA of a frog. Wolves and frogs are very different and are thought to have diverged from a common ancestor many millions of years ago. There should be more differences in DNA between a frog and a wolf than between a fox and a wolf. And that is what is found. According to evolutionists, this has occurred because there has been more time and more generations since a wolf and frog had a common ancestor for mutations to accumulate in each line of descent independently than the time since the wolf and the fox had a common ancestor.

The differences between the DNA of various species seems to confirm common descent. There is a close match between DNA differences and differences in body structure, body chemistry, and behavior. These differences can be explained by inherited changes due to mutation from a common ancestor over time.

Those who believe in intelligent design have an answer, and that is that the differences in DNA are not due to random mutations and natural selection but due to intelligent design. To produce different body structures, body chemistry, and behavior, an intelligent designer coded different DNA to produce these outward changes.

Ok, there is truth in that, but also problems. Not all differences in DNA necessarily produce differences in the characteristics of an organism, and those that do not are therefore not explained by design. Some changes in DNA seem to have little or no effect on the characteristics of an organism (as far as science has so far discovered).

Not all DNA is in genes.

A review of the structure and functioning of DNA will help. This will be familiar to many readers who have studied biology.

Genetic material that determines the characteristics of an organism is contained in chromosomes, and those chromosomes exist in the nucleus of a cell. Different species have different numbers of chromosomes, and humans have 46 chromosomes, 23 from the mother and 23 from the father.

Inside each chromosome is a long strand of DNA. The DNA is made up of a series of nucleotides. Each nucleotide has a base, and there are four different base molecules—adenine, thymine, cytosine and guanine. They are known by their abbreviations—A, T, C, and G.

They can be in any order in the DNA chain. They are therefore free to code information—information that determines the characteristics of an organism.

They do this by specifying amino acids, and the amino acids are used to build proteins. The proteins in turn are the building blocks that help make up the organism and determine its characteristics. It is thus the order of the DNA bases that determines the proteins in an organism and thus its characteristics.

A sequence of three DNA bases make a DNA base triplet, also called a "codon." I will use the term, "DNA base triplet" or just "triplet." Since there are four possible DNA bases, there are 64 possible triplets. The first base in the triplet can have one of four values. For each of those values, you can have any of four possible bases in the second position in the triplet, so with two bases you have 4 X 4 or 16 possible combinations. Then, in the third position you can have 4 possible bases, so 16 X 4 = 64. Thus there are 64 possible DNA base triplets.

The purpose of a DNA base triplet is to specify a particular amino acid to be used to build a protein molecule.

Although there may be hundreds of amino acids possible in chemistry, only twenty particular amino acids are used in life to make proteins. Each DNA base triplet can specify a particular amino acid. These amino acids are strung together, in the same sequence as the DNA base triplets that specify them, into long chains. These chains of amino acids are the protein molecules. Each chain folds into a particular shape, the shape of the protein molecule, and that protein molecule functions in the cell to produce the characteristics of the organism. Millions of kinds of proteins are possible.

But notice, there are 64 possible DNA base triplets to specify 20 possible amino acids. An amino acid can be specified by more than one triplet. So there is redundancy in the DNA code. Several different triplets can specify the same amino acid. There are also a few triplets that are used to stop or start a sequence. They are like punctuation—like a period at the end of this sentence.

A gene is a particular sequence of DNA base triplets that code for a particular sequence of amino acids that make up one protein. Thus a gene specifies a protein. The gene has a beginning, a middle, and an end.

But not all the DNA in a chromosome makes up functional genes. There are long sequences of DNA bases between genes that are not functional genes. Some of these are pseudogenes, dead genes that, because of mutations to their coding, cannot function as genes and have no effect on the characteristics of an organism. They cannot produce proteins. There is also noncoding DNA, what some call "junk DNA"—DNA between genes that does not make up genes at all and in its sequence of bases does not seem to have the same strong and direct effect on the characteristics of an organism as the sequence of bases in genes does. These sequences are not genes and they do not produce proteins.

The pseudogenes and noncoding or junk DNA may actually make up more DNA than the working genes in many or most organisms, including humans.

But this may be controversial. Some scientists may claim to have found that much of the DNA outside of protein-producing genes does have a function, though it does not produce proteins.

Besides the DNA inside the chromosomes that are in the nucleus of a cell, there is DNA in mitochondria—structures that are involved in energy processing—scattered in the cell. This mitochondrial DNA is generally passed from the mother to the offspring, but not from the father, in contrast to DNA in the chromosomes that are passed from both mother and father to offspring. This mitochondrial DNA does not, in its sequence of bases, seem to have a major determining effect in the characteristics of the organism.

Now let's once again look at the molecular clock as evidence of common descent.

Some may say that differences in DNA between species are solely due to a designer coding the DNA to produce the desired outward characteristics of an organism. But there seems to be a pattern of differences and similarities of DNA between species that seems to indicate common descent, not intelligent design, because this pattern appears in several kinds of DNA that do not seem to determine the characteristics of an organism, and therefore there is no reason for a designer to code such a pattern.

There are several kinds of DNA that do not have any major effect on the characteristics of an organism that scientists have discovered so far.

Mitochondrial DNA is outside the nucleus of a cell, and though it is inherited through the mother, it does not seem to have any major effect on the characteristics of an organism as genes do.

Pseudogenes do not function as genes, yet are part of the nuclear DNA that is passed from one generation to the next. The DNA coding in the pseudogenes does not affect the characteristics of the organism.

So-called junk DNA makes up a portion of nuclear DNA that is passed from one generation to the next, yet the sequences of bases in this DNA do not code for proteins that determine the characteristics of an organism.

Intelligent design has difficulty accounting for the pattern of differences and similarities in base sequences between species in the above categories, a pattern that is easily explained by random mutation and common descent over millions of years.

There is also redundancy in the coding of DNA base triplets to produce the amino acid sequences that make proteins. If a random mutation changes a DNA base triplet to a different triplet that codes for the same amino acid as it did before, that mutation will not specify a different amino acid in the protein it codes for and thus cannot easily be explained by design. Yet that mutation will be passed on from generation to generation and can be detected in species today.

For example, you might have a DNA base triplet of AAA (adenine, adenine, adenine) in one of a wolf's genes sometime in the past. This gene may be the same in a fox. This triplet codes for the amino acid lysine, which will be part of a particular protein. Then a mutation changes the triplet in the wolf to AAG. That change will be passed on to future generations in the wolf, but not the fox. But the different triplet, AAG, *also* codes for the same amino acid, lysine. It makes the same protein, so it has no effect on the organism. So design advocates cannot say that this difference was designed to produce changes in the organism. The only obvious explanation is random mutation and common descent.

When a wolf and a fox have a smaller number of such differences than a wolf and a frog, such a pattern cannot be easily explained by design, but can be explained by common descent. There are more differences between a wolf and a frog than between a wolf and a fox, because there have been more generations and more time between a wolf and its common ancestor with a frog than between a wolf and its common ancestor with a fox. When the wolf and the fox first branched off from a common ancestor, their DNA would be the same. But as time went on, differences due to mutation would accumulate in each line, and the longer the period of time and the greater the number of generations since they branched off to become independent lines, the greater the number of differences will be.

Mutations change DNA in the cells, and then those changes are passed on from generation to generation. These changes occur over time, and they can be used to identify which species are closely related and which are distantly related. The more DNA differences there are between two species, the longer the period of time since they had a common ancestor. This pattern matches up fairly well with the pattern of outward characteristics of organisms, such as body structure. This can be explained by common descent. The pattern of similarities and differences between species in the DNA contained in active genes can also be explained by design, but not necessarily the same pattern in pseudogenes, noncoding or junk DNA, redundant DNA base triplet coding for amino acids, and mitochondrial DNA.

There is also a pattern of resemblance in the order of genes and DNA sequences in the chromosomes which do not affect the characteristics of the organism.

Nonfunctional DNA sequences, which are independent of any observed characteristics of an organism such as structure, chemistry, and behavior, show the same structured pattern of resemblance as body structure and other characteristics, but cannot readily be explained by intelligent design. But they are readily explained by common descent and natural mutation over time.

Thus, much of the pattern of similarity and differences in DNA structure between species seems to be neutral as far as the characteristics of the organisms are concerned and is not easily explained by design. Yet, they show a pattern that can be readily explained by common descent.

To me, this is a very strong argument in favor of species having a common ancestor and emerging by common descent.

There is also the fossil record. Scientists can estimate the approximate age of various strata in the earth. Although they can sometimes do this by fossils themselves, determining how old a strata of earth is by the estimated age of the species of fossils in the strata, they have other means of determining age of strata without using fossils. Radiometric or radioactive dating can be used to determine how old rocks are in the strata, and this is based on the rate of decay of radioactive elements, which is based on fundamental laws of physics and does not change. There are also layers of sediment in the strata, and these layers can show which strata was laid down first and which last. And although strata can be overturned, there are overlapping layers that can be matched up in different parts of the world to show a consistent pattern.

And scientists find that fossils of simpler species appear in older strata, and more complex species appear in younger strata. Also, fossils of those species similar to today's species appear in younger strata. This also supports the idea of the emergence of species through common descent.

Finally, there is the existence of vestigial organs in species today, organs that seemed based on a function that existed millions of years ago but not today, such as leg bones in whales. This also suggests common descent and gradual change.

# Emergence of Species Gradually over Millions of Years

The two pieces of evidence that support the idea that species emerged over millions of years are two used to support common descent: DNA and the fossil record.

It takes time for many mutations to appear in DNA. Scientists can estimate how long it may take for mutations to change the DNA based on examination of existing species and variations within those species, and it would take millions of years for the changes to appear between dissimilar species. Again, many of these differences appear to be due entirely to random mutation, not design, because major portions of DNA do not seem to determine the characteristics of an organism.

The fossil record also shows that species appeared gradually over millions of years, because it took millions of years for the normal geological processes to produce those strata.

When I say that the species emerged gradually, I do not mean that every individual species gradually emerged from a parent species. Some species may have emerged rather abruptly from a parent species, and the theory of evolution has a difficult time explaining it. Transitional forms are often missing.

But when I say life forms appeared gradually over millions of years, I am talking about the entire biological family tree. It appears to have taken millions of years, even hundreds of millions of years, for the fossil record to be laid down.

Scientists can date strata in the earth by various radiometric or radioactive dating methods. Radioactive elements decay into other elements or isotopes of elements, and they do so at a certain fixed rate depending on the particular element or isotope. These rates are determined by fundamental laws of physics that do not change. Scientists can measure the proportions of decay elements with the original radioactive element to estimate the age of a rock, and this can be used to determine the age of the strata. I do not claim that these results are one hundred percent accurate in detail, and mistakes can be made, but overall these measurements are sufficient to establish that the earth and various strata are millions of years old, and these methods can be used to determine the relative sequence of the strata—which strata is oldest, which is youngest, etc.

It turns out that fossils of simpler life forms—life forms that would be expected based on structure to be ancestral to later life forms—are in the older strata, while life forms that are more complex and appear to have descended from earlier ones are in younger strata.

The existence of radioactive elements and their decay elements and isotopes, and the rate of that decay, which is established by the laws of physics that do not change, show the great age of the earth and the universe.

Thus you have patterns of resemblance in body structure, DNA, and age of fossils that point to a consistent picture, explained by common descent over millions of years but not perfect and sudden creation by an infinite Creator six thousand years ago.

There is also evidence, such as the cosmic background radiation, that shows that the whole universe is billions of years old.

## Would God Create False Physical Evidence?

Some who believe in God have suggested that God made the evidence in the earth to make it *appear* as if the earth were older than it was, perhaps to test our faith, but that the apparent age of the earth and life and common descent that this evidence seems to point to is not real. They may use the examples of tree rings or Adam's navel. They claim that the trees that first existed in Adam's day had tree rings giving the appearance of age and that Adam had a navel as if he had been born.

Regarding the fossils, some may have suggested that God allowed Satan to create false fossils of creatures that never existed to deceive us, and God allowed it to test us.

Some have used this argument even regarding our observations of the universe itself.

But God would not create false evidence of fossils, radioactive elements, and DNA structure to deceive us, and I do not believe He would allow Satan to do so either. God does allow Satan to create false signs and wonders, but God warns us about this first in His word, the Bible (Matthew 24:24–25; Revelation 13:11–15).

God teaches us not just by His word but by experience and observation, including observation of the natural world. In fact, it is by observation of the natural world that we find proof of God's existence. Indeed, God's word confirms that we should know He exists by the things we see. "For the wrath of God is revealed from heaven against all ungodliness and unrighteousness of men, who suppress the truth in unrighteousness, because what may be known of God is manifest in them, for God has shown it to them. For since the creation of the world His invisible attributes are clearly seen, being understood by the things that are made, even His eternal power and Godhead, so that they are without excuse" (Romans 1:18–20). Christ even showed Thomas evidence that He, Christ, who was crucified, was raised alive and was present so Thomas could believe. "Then He said to Thomas, 'Reach your finger here, and look at My hands; and reach your hand here, and put it into My side. Do not be unbelieving, but believing' " (John 20:27).

Since God uses creation to communicate His existence, He cannot deceive us with deceptive tricks in His creation. See also Psalm 19:1–4 that shows that God speaks through His creation. Since God has intended that creation make His existence known, why would He create a deceptive picture that makes it look like the universe is older than it is? God does not just communicate with us through His word but through creation also.

Since I mentioned it, I cannot resist giving my opinion about tree rings and a navel after Adam was made. I think God would not make these features just to make things look older than they are. If they serve a structural purpose, yes, otherwise, no. Do tree rings add structural strength to a tree? If so, God probably made tree rings to make the trees strong. If not, then I see no reason why God would make rings in the trunks of the trees He made directly. Does a navel serve a purpose for one not born? I see no purpose, so unless God had a purpose we don't know about, my guess is that Adam probably did not have a navel.

God would not create false evidence of age or common descent in the earth, in the fossils, in the DNA, or in the universe, nor would He allow Satan to do so. The evidence is true and available for us to study.

## Emergence of Species Due to Natural Forces Only

The third major part of the theory of evolution is the belief that all changes that took place to produce any new species from a parent species are due to natural forces only, not intelligent design. The mechanism Darwin proposed was random mutation (or variation) and natural selection. Mutations would occur in the nuclear DNA of some organisms (Darwin did not know about DNA, just variation, not knowing its source and mechanism), changing the characteristics of the offspring of those organisms. When those changes were beneficial, by chance, the new organism would be successful and reproduce, thus spreading the change. A number of changes of this kind would eventually result in a new species.

Although some changes can occur this way, the evidence that this is the only way changes occur is nonexistent. Evolution and the science community that promotes evolution offers no argument to exclude design and intervention by God or any other intelligent designer, except their blind faith that there is no God who intervenes in the natural world. In fact, there is evidence that natural forces cannot explain all the species.

I already mentioned the lack of transitional forms between species in the fossil record. This is a problem for evolution. If each species only emerged gradually with few beneficial mutations per generation, you would expect many transitional forms from one species to the next in the fossil record. But those transitional forms are missing.

But intelligent design easily explains the lack of transitional forms. With intelligent design, the genetic changes needed to make a new species from a parent species could be designed and made at once in one generation—there would be no transitional forms.

There is also the problem of irreducible complexity. That is a fancy term to describe the situation where you need many things working together before any one of them is useful for survival or reproduction. The odds against beneficial mutations occurring all at once to bring *all* of these characteristics into existence in one generation are astronomical. Yet, if one or two came into existence, they would provide no reproductive advantage, and they would die out.

There are many examples of this in the natural world. One of the most powerful is the DNA/protein mechanism. In the cell, DNA is needed to make protein. It is the sequence of DNA base triplets that determines the sequence of amino acids that make a protein. Yet, proteins are needed to process the DNA code to produce proteins. You need both at the same time to function in a living cell. How could this have evolved in steps due to natural causes only? Science has no answer.

Bees and other pollinating insects need flowering plants to obtain their food from pollen. But flowering plants need pollinating insects to reproduce. How could they evolve together only through random mutation and natural selection? An insect that would pollinate a flower would have no reproductive advantage for doing so if flowers did not exist, and plants would have no reproductive advantage to flowers if pollinating insects did not exist.

Next time you see a monarch butterfly on a flower, think about what came first, flowers that need insects to pollinate them or insects that need the nectar or pollen for food.

Or take the example of metamorphosis where a worm or caterpillar changes into a flying insect such as a fly or butterfly. How could that possibly have evolved in gradual steps through random mutation and natural selection only? Something else to think about next time you see a butterfly.

The woodpecker is another example. A woodpecker feeds off ants that have tunneled into trees and tree limbs. It has an especially hard beak for piercing wood. It has especially strong neck muscles for repeatedly hammering the wood. It has especially acute hearing to locate the ants by sound. It has a special tail to brace itself as it hammers the wood. These and other tools have to work together. The odds against all these things happening as the result of random mutation in one generation are astronomical, yet if any one happened, it would soon die out because it would be of no advantage by itself. If, for example, a mutation resulted in an especially hard beak, without the strong neck muscles, instincts, and other things needed to hunt ants in trees, not only would it do no good, but it would be a disadvantage, since there would be a biological cost for a species to maintain especially hard beaks that are not needed.

Evolutionary theory may have difficulty explaining all these things, but intelligent design has no difficulty. In the case of the woodpecker, an intelligent designer could design, in one generation, all the changes needed to produce the species from a parent species.

There are many examples of irreducible complexity, and I refer the reader to the bibliography for books that give more details than I have room for here.

Evolutionists try to explain how some examples claimed to be irreducibly complex could have occurred in steps, and some cases they may be able to explain, but not everything. Perhaps even the case of the woodpecker can be explained in steps—I offer this example in detail because it is a good illustration that teaches the concept. But there are other examples more difficult to explain.

There is also the problem in the fossil record with missing transitional forms and rapid development of new life forms. Although the fossil record shows gradual emergence of life forms in the larger sense, it is a puzzle why more transitional forms have not been found between different life forms and why new forms seem to appear more suddenly than expected. The Cambrian explosion, when many new forms appeared in a relatively short amount of time, is an example.

The evidence that atheists and majority science claim points to evolution is really evidence for only two parts of the theory of evolution—common descent and gradual development over millions of years of many life forms and species—but not through natural forces only.

But why would God, who is assumed to be the designer, do this over millions of years through common descent? If evidence shows that species emerged gradually over millions of years from common ancestors, why would God design and create the species of life that way? God is infinite in knowledge, wisdom, and power. He could create all the species of life immediately—in one generation—without common descent, without millions of years in time.

What did God actually do, and why did He do it the way He did?

That is the subject of this book. But before I offer an explanation, let's review the evidence of science and the Bible and see if there is a contradiction. And before that, let's see what the evidence says about God's existence.

There is evidence that there is a God, though atheists do not acknowledge it. But this is the place to review it.

## Does Evidence Point to a Creator God?

The universe and its laws seem to be fine-tuned for man's existence. This requires an explanation.

I know of only two explanations that offer any degree of plausibility. One is that God exists and He designed and created the universe and its laws in a way that makes man's existence possible. The other is that all possible universes exist (the "multiverse" idea or one of its variations) and we find ourselves, naturally, in one of those universes that allows us to exist. The anthropic principle is involved here.

The problem with the multiverse idea is that the universe is not just fine-tuned for our existence. It is fine-tuned for our observation of the universe. If there is an infinite number of universes, every logically possible universe, it is inevitable that we would find ourselves in one that is fine-tuned for our existence, since we would not otherwise be around to debate the question. But it is not inevitable that we would find ourselves in a universe fine-tuned for our observation of the universe.

But the fine-tuning of the universe for our observation of the universe can be explained by the design of the universe by a creator God who wants us to see and know how vast and elegant the universe is and how great is the God who made it. He fine-tuned it for discovery.

Consider the apparent size of the moon. The moon, though much smaller than the sun, is much closer and thus appears to be the same size in the sky. But the moon was not always the same distance from the earth as it is today. In times past, the moon was closer. Due to tidal forces, it has been moving farther from the earth since the earth-moon system existed. Millions of years ago, it would have appeared larger than the sun in the sky, and millions of years from now, if the earth-moon system continues to exist and function as it does today, the moon will appear smaller than the sun.

Atheists say that man evolved from natural causes over the last few hundreds of thousands of years, or at most a few million.

Why did man appear on earth with his intellect, his inquiring mind, his powers of observation just at the same time that the moon was the exact distance from the earth for it to appear as the same size as the sun? Is this just a coincidence? Or did God design it that way? And if God designed it, why? Perhaps for esthetic reasons only?

While esthetic reasons should not be excluded, there is another possible reason. It is the apparent size of the moon, just large enough to blot out the sun's disk in the sky, but not the stars that appear in the sky around the sun, that helps to confirm Einstein's general theory of relativity. That theory helps us observe and understand the universe as a whole. It helps to show God's greatness and humbles us to know that the greatest scientific minds cannot reconcile general relativity with quantum mechanics, though observation and experiments confirm both.

We also find ourselves in a position in the galaxy between spiral arms of the galaxy that give us a clear view of the universe unblocked by a concentration of stars too heavy to see through. There can be reasons why the requirements of life can restrict our location to be between the arms of the galaxy, but if all possible universes exist, there would be an infinity of universes with laws that allow our existence but do not require us to be able to observe the universe.

Both general relativity and our observation of the distant universe help us to know that the universe had a beginning, and this beginning points to a creator God.

These are just a couple of examples of how the universe is fine-tuned, not just for our existence, but for our observation of the universe. The multiverse idea and the anthropic principle may explain, to some people, our existence apart from God, but it cannot explain the fine-tuning of the universe for our observation of the universe. If the multiverse idea were true, we could just as easily exist in a universe fine-tuned for our existence but not our observation of the universe. We would be around to debate God's existence, but without the knowledge of science and the universe we have.

It is as if God wants us to see the magnificence of the universe He has created. "The heavens declare the glory of God; And the firmament shows His handiwork. Day unto day utters speech, And night unto night reveals knowledge. There is no speech nor language Where their voice is not heard. Their line has gone out through all the earth, And their words to the end of the world" (Psalm 19:1–4).

Specifically, the universe is fine-tuned for us to observe its immensity and complexity and also *that it had a beginning*. Both of these reveal the existence and infinite power and wisdom of God.

The multiverse idea and anthropic principle cannot explain the fine-tuning of the universe for our observation and knowledge of the universe, but God's design and creation of the universe can.

I might point out that the multiverse idea can never disprove God's existence in any case, for if every possible universe exists, then there would be a universe in which God exists.

One can also ask, if there is an infinity of all possible universes, why do we not live in a universe where intelligent life is common? But it is, in fact, rare, so rare that we find no evidence of biological intelligence except on earth.

There can be versions of multiverses where the number of universes is large but not infinite and not every kind of universe exists, but that brings us back to a problem. Some designer would have to design what kinds of universes exist and what kinds do not.

This universe does not have to be the way that it is. Design choices have been made, in the number of dimensions, the number and characteristics of the forces of matter and energy, the kinds and nature of particles that exist, etc. These design choices point to a designer who designed the universe for our existence and observation. He made the universe so that anyone with an open mind can find evidence of His existence and power, but without forcing those who do not want to believe in Him to acknowledge His existence. He gives people a choice at this time, to believe in Him or not. But for those whose minds are open, He gives proof that He is.

There are many examples in nature and in the laws of the universe of fine-tuning both for our existence and our observation of the universe and the discovery of the laws of nature. The reader can do research with the books I have listed in the bibliography at the end of this book.

Add to the fine-tuning of the universe the evidence of the irreducible complexity of life in all its varieties, and I find a convincing picture that God exists and designed and created the universe and life on earth.

But there is one more piece of evidence, and this deserves a separate section.

Human consciousness.

## The Mystery of Consciousness

What is consciousness? David Chalmers in his writing and speaking explains it better than most. I'll give my own definition here.

I would describe consciousness as a subjective sense of experience we have apart from any mechanical functioning of the machine we call the human body and brain. It is the "I" that exists in my brain. It is the part of me that suffers with pain or enjoys the experience of happiness.

Many scientists have trouble even defining it or describing it. They tend to confuse it with intelligence, with "self-awareness" as when a person (or animal) looks in a mirror, or attention, or some part of the brain that lights up in a brain scan when we are doing a certain mental task. But it is none of those things.

Our brains are like machines made up of a complex arrangement of protons, neutrons, and electrons, operating according the the laws of physics, chemistry, and biology. They are like computers, only far more advanced than any computer man has made. They process information in incredibly complex ways.

Scientists and doctors have learned a lot about the human brain. They know what parts of the brain handle what parts of mental processing, from interpretation of visual signals from the eyes to motor control over our arms and legs and hands.

Animals also have brains that process information.

The intelligence that we humans and animals have as a result of brain structure and processing is necessary and helpful for survival and reproduction. And much of this processing can be explained by scientists as a result of physical matter and energy operating according to the physical laws of nature. No doubt, scientists will learn much more as they continue their studies.

But none of this has anything to do with consciousness.

Consciousness is a total mystery to science. Each of us knows we have it because we experience it directly. We assume other humans have it because we think we are the same in that regard and because other humans can discuss consciousness as each of us can. Many of us assume animals are conscious.

But it plays no role, itself, in survival and reproduction (intelligence does that), and it cannot be explained as a function of the laws of physics. Yet it exists.

There is no reason for it to evolve. And there is *no way* it could evolve.

Many scientists do not like to talk about consciousness because it threatens their materialism.

Creation is the proof of God. God inspired Paul to write this in Romans 1:18–20 in the Bible.

Atheists will counter the evidence of the fine-tuning of the universe with their argument that the universe with its matter, energy, and laws has always existed in some form or is only one of every possible universe. They will construct some theories, not entirely rational, about how this is possible. They use the theory of evolution to try to explain life in its varieties.

But none of this works for consciousness. It is a separate thing altogether.

Atheists claim evolution and materialism can explain life and a multiverse with the anthropic principle can explain the universe, but neither explains consciousness. Consciousness is a part of creation that points to a creator that science and materialism can't touch. Consciousness has no useful function for survival and reproduction and can't be explained by materialistic evolution.

My conscious mind exists. It did not exist a hundred years ago. It came into existence. How? The only explanation I find plausible is that God made my mind. Yes, I was born through normal biological processes. My body and brain grew through natural forces. But God must have made my consciousness, for the laws of matter and energy do not apply here. This, for me, is a proof of God. I find no other explanation plausible.

## Other Evidence that God Exists

There is proof in the Bible and history that the Bible was inspired by a God who was able to know and predict the future. Fulfilled prophecy is a proof of the inspiration of the Bible and a proof of God. There is also internal evidence in the Bible of the inspiration of God. I explained some of this in chapter two.

Also, there are many people who have personally experienced God in their lives in a way that cannot be plausibly explained as coincidence. People have received answers to prayers and have seen God intervene in their lives in various ways. These instances are sometimes of a highly personal nature, and I suppose no person could convince another person of God's existence by relating such incidences, for the listener could always conclude that the speaker was not telling the truth or was misinterpreting what happened.

But for those who have experienced God's intervention in their lives and understand it, this is certainly evidence for His existence.

## Summary

God shows enough evidence of His existence to allow anyone with an open mind, anyone *willing* to believe in Him, to prove that He exists. But He does not show so much evidence that those who do not want to believe in Him are forced, against their will, to believe He exists. He gives people a free choice about that at this time. He won't force you to believe in Him. But if you are willing to believe the truth, whatever it is, He gives evidence of His existence. Creation is the main part of that evidence, but not the only part (there are other kinds of evidence, such as fulfilled prophecy in the Bible, answered prayer, miracles and interventions in people's lives, etc.).

Physical evidence seems to show that the species of life came into existence over millions of years from a common ancestor. There is evidence of an old earth and an old universe, and I have only covered a part of that evidence here.

But there is also evidence that life and the species of life could not have come into existence through natural forces only. Random mutation and natural selection could not have produced, by themselves, the examples of irreducible complexity that exist in the natural world.

And there is abundant evidence that God exists, though at this time He allows man to choose to deny His existence if they want.

Yet, there is still a mystery.

If God exists and has designed and made the universe and made the species of life, why would He take millions of years to do it? Why would He use descent with modification from a common ancestor? Why would He do things incrementally, building more complex organisms over vast amounts of time. Why would He make imperfect organisms with vestigial organs?

And why would a good, loving, and righteous God design a competitive natural world with animals killing animals, sometimes in very cruel ways? We see that in the world today and in the fossil record. Some of the dinosaurs perfectly illustrate the competitive and cruel nature of the natural world—T. rex comes to mind.

In a larger sense, one could ask, why would an infinitely powerful and righteous God allow all the suffering and injustice in the world today?

If you look at the physical evidence in fossils, in the rocks and in strata, in the structure, behavior, and chemistry of living species today, and in DNA, what you find is something puzzling and amazing.

You find evidence of intelligent design, but *limited* and *imperfect* design, sometimes even cruel and unrighteous design. You find design by an intelligence that is powerful but limited and imperfect.

This is what people on both sides of the evolution vs. creation debate miss. Atheists do not want to believe in any intelligent design, perfect or not, because to them intelligent design means God and the supernatural, which they choose to deny, for personal reasons in many cases. Bible believers do not want to believe in an old earth because they mistakenly think the Bible teaches a six-thousand year old earth, and because they assume God suddenly created all the species (or major "kinds" of life) individually and directly, using His infinite intelligence, wisdom, and righteousness to make a perfect creation all at once.

So what is the answer?

I will suggest an answer in the rest of this book, one that is consistent with all the physical evidence and is consistent with a literal understanding of Genesis and the Bible.

# CHAPTER FOUR

# Does Scientific Evidence Contradict the First Chapter of Genesis?

Science and the scientific community dominated by atheists do not accept the account of creation in the Bible as given in the first chapter of Genesis. They claim it is a myth.

The traditional Christian community, not wanting to reject science and scientific knowledge and physical evidence altogether, but wanting to retain some kind of faith and belief in the Bible, has proposed various ways of reconciling Genesis with science.

## The Young Earth Belief

One way is simply to reject scientific claims for an ancient earth and claim that the earth itself is only six thousand years old. This may be called a belief in a young earth.

There are a couple of problems with this. One is that there is good scientific evidence that the earth is older than six thousand years. Those who are knowledgeable about science would have difficulty accepting this idea. Likewise, students who are taught geology and science would not be able to explain how the earth could only be six thousand years old.

Another problem is the dinosaurs. There is an abundance of fossil evidence indicating that dinosaurs, which do not appear today, existed at some time on the earth. If the earth did not exist before the six days of creation as described in Genesis, then they must have been made during the six days, or after. But before the flood in the days of Noah, according to Genesis 6:13–22, *all* the air-breathing land animals were brought into the ark. Yet, dinosaurs do not exist today, and there is no record in scripture of any mass extinction after the flood. Moreover, even with the large size of the ark, it is difficult to imagine how some of the largest dinosaurs would fit (some have suggested that they would fit as babies).

Moreover, we have already seen in chapter two that the Bible does not say that the earth is only six thousand years old. Genesis and the Bible allow for an earth that is hundreds of millions or even several billion years old. The earth was created and existed *before* the six days— perhaps billions of years before. Then the surface of the earth was destroyed, probably as a result of the sins of the angels, just as it was destroyed by a flood because of the sins of mankind in the days of Noah. Then, God renewed, refreshed, restored the surface of the earth in six days.

The dinosaurs apparently existed and became extinct millions of years before the six days in Genesis 1. We see their fossils today.

## Belief in a Figurative Interpretation of Genesis

Many traditional Christians accept the Genesis account as symbolically or figuratively true but not literally true. They say that God did not intend that it be taken literally. According to them, the six days represent general periods of time, each of which could have lasted millions or hundreds of millions of years.

But there is nothing in the account of the six days that indicates that figurative language is being used. It sounds quite literal and would have been understood in a literal sense by people who have read it for thousands of years until recent times.

And as I pointed out in chapter two, God, if we are to trust Him and His word the Bible and have faith that God cannot lie and will only tell us the truth (Titus 1:2; Hebrews 6:18; John 10:35), cannot tell the human race and His church through the ages something He knows they will take literally for hundreds of years, then later say, "What I told you is not literally true—I was only speaking figuratively." He would not deceive us that way. If there is no evidence in the Bible that God is speaking figuratively, then He must be telling us the truth literally.

Some may say that ancient people could not understand modern science, and so God had to give them an explanation they could understand, which they say the Genesis account is when taken figuratively.

But that makes no sense. God did not have to build an elaborate mythology of six days of creation just because ancient peoples did not understand modern science. Whatever God did, He could have given a simple explanation they could understand that is still literally true. God could simply have stated that He made the earth and all life on it and let it go at that. And if He had said that in the Bible, we would be free to speculate if God used the process of evolution to develop life right up to the present time.

But that is not what the Bible says.

The six days of creation really happened, literally. If we accept the Bible, if we have faith that God inspired the Bible and will never lie to us, if we have faith that God's word, scripture, cannot be broken (John 10:35), then we must understand the six days of creation literally.

If God left us free to interpret figuratively any portion of scripture we do not understand or do not want to agree with, we could not rely on *any* part of the Bible. We could say that none of it is literally true, and thus, none of it would be reliable. Each person could interpret the Bible his own way, picking and choosing what to accept and believe literally and what to dismiss as figurative language to mean whatever the reader wants to interpret it to mean.

And many do just that. But that is not what God intends for those who love, believe, and obey Him.

"Thus says the LORD:
'Heaven is My throne,
And earth is My footstool.
Where is the house that you will build Me?
And where is the place of My rest?
For all those things My hand has made,
And all those things exist,'
Says the LORD.
'But on this one will I look:
On him who is poor and of a contrite spirit,
And who trembles at My word'" (Isaiah 66:1–2).

"The entirety of Your word is truth,
And every one of Your righteous judgments endures forever" (Psalm 119:160).

"Sanctify them by Your truth. Your word is truth" (John 17:17).

We cannot have the respect and awe for God's word and its authority over our lives that we should have if we cannot trust what God says.

## The Gap Teaching

The so-called "gap theory" or gap doctrine is the idea that Genesis chapter 1, verse 1 describes the original creation of the earth, which could have occurred millions of years ago, that the surface of the earth was destroyed resulting in the "without form, and void" condition as described in verse 2, and then, starting in verse 3, God began the process of renewing the face of the earth and restoring life to the earth in six days.

I talked about this in Chapter Two. This actually seems to be what has occurred. This understanding allows for the existence of dinosaurs and other plants and animals we see in the fossil record but today are extinct, and it allows for scientific evidence of a very old earth. It does not contradict the physical evidence science had discovered, and it is faithful to a literal understanding of the Bible account.

But this approach leaves a lot of unanswered questions—questions Christians find hard to answer.

If physical evidence of fossils and DNA really point to the gradual emergence of the species over millions of years through common descent, why would God take so long to make all the species of life? Why would He do it that way? God is all wise and all powerful. He could design and create all the species of life in one day, with each species being an individual creation not descended from other species. Why do it through common descent over millions of years?

Why the dinosaurs? Why would God create so many dinosaur species just to bring them to extinction later? Why would there be a mass extinction of many species millions of years ago?

And why did God create predator animals that kill and eat other animals? Is this what God intended from the beginning?

The Bible indicates that Christ will return to this earth to set up the kingdom of God, and Christ and the saints will rule the earth for a thousand years. "And I saw thrones, and they sat on them, and judgment was committed to them. Then I saw the souls of those who had been beheaded for their witness to Jesus and for the word of God, who had not worshiped the beast or his image, and had not received his mark on their foreheads or on their hands. And they lived and reigned with Christ for a thousand years. But the rest of the dead did not live again until the thousand years were finished. This is the first resurrection. Blessed and holy is he who has part in the first resurrection. Over such the second death has no power, but they shall be priests of God and of Christ, and shall reign with Him a thousand years" (Revelation 20:4–6).

During that time, the nature of animals will be changed.

"The wolf also shall dwell with the lamb,
The leopard shall lie down with the young goat,
The calf and the young lion and the fatling together;
And a little child shall lead them.
The cow and the bear shall graze;
Their young ones shall lie down together;
And the lion shall eat straw like the ox.
The nursing child shall play by the cobra's hole,

And the weaned child shall put his hand in the viper's
     den.
They shall not hurt nor destroy in all My holy mountain,
For the earth shall be full of the knowledge of the LORD
As the waters cover the sea" (Isaiah 11:6–9).

   " 'The wolf and the lamb shall feed together,
The lion shall eat straw like the ox,
And dust shall be the serpent's food.
They shall not hurt nor destroy in all My holy
     mountain,'
Says the LORD" (Isaiah 65:25).

   No more will animals hunt and kill and destroy other
animals. Why? It is evident that the competitive cruelty
in the animal world is not God's intent for the millennium when the knowledge of God will fill the earth and
Christ will rule the world in righteousness, peace, and
happiness.

   So in the millennium, the thousand-year reign of
Christ and the saints, there will be peace all over the
earth.

   "He shall judge between the nations,
And rebuke many people;
They shall beat their swords into plowshares,
And their spears into pruning hooks;
Nation shall not lift up sword against nation,
Neither shall they learn war anymore" (Isaiah 2:4).

   Even the nature of wild animals will be changed.
Wolves and lions will be peaceful and will not hurt or
destroy other animals.

But if that is God's will for the millennium—if that is an example of the perfect peace even in the animal kingdom that God intends—why would He create predatory animals that kill and destroy other animals in the first place? And why T. rex, one of the greatest killer dinosaurs of them all?

This point is actually a strong reason why many people have accepted evolution, especially in the early days of evolutionary theory. I think Darwin himself was concerned about it. If God created the species, and if God is good, why would He create such a competitive system in the animal kingdom where animals selfishly hurt and kill and destroy other animals?

Those are questions I will try to answer in the chapter that follows.

## Noah and the Ark

Before leaving this chapter on Genesis, I want to briefly address the topic of the flood in Noah's day and the ark.

Questions sometimes come up about the practical aspects of saving the air-breathing animals in the ark and the account of this in Genesis. Critics take issue with one thing or another, looking for details of problems, real or imagined, and claiming that the Genesis account of the flood and the ark is impossible. But they are not looking for solutions.

But solutions are possible for those who look for them. Female animals on the ark could have come aboard already pregnant from a different male that accompanied them, adding to the genetic diversity of the species. Large animals could have come aboard as babies, saving space and food. Many problems disappear when solutions are sought for, and this is not the place to go into a thorough study of the flood and Noah's ark.

Some may wonder how the different animals from the ark got to the different parts of the world where they belong after the flood—Australian animals to Australia for example.

One might also wonder how those animals from Australia and every part of the earth got to the ark before the flood in the first place.

But there is an important principle to keep in mind.

The flood and the saving of Noah and his family and the animals to repopulate the earth was God's act, and He has all power and wisdom. Nothing is impossible for Him. He could do any miracles required for any reason at any time.

God had Noah build an ark to house the animals because God wanted Noah to have a part in this event. God is teaching lessons, and the flood and the ark and Noah's work were part of the lessons God teaches us. So God gave Noah and his family a job to do, and whatever they could not do, God was able to work out miraculously.

God was in charge, and He did whatever Noah was not able to do. God worked miracles to bring those species of animals into existence during the six days of creation week. He worked a miracle to bring on the flood.

Noah did not have to go to different parts of the world to gather those animals before the flood. He did not have to search out and capture and bring into the ark representative samples of every species from every part of the earth, nor did he have to return those animals to their native lands. God *brought* them to Noah and the area of the ark, working whatever miracles were necessary to do this. And the same miracle-working God was able to take those animals back to their home areas after the flood by whatever miracles were necessary for this purpose.

Actually, God could have done everything Himself, preserving a number of animals alive during the flood by miracles without Noah's help and without an ark.

But it suited God's purpose to give mankind, in the person of righteous Noah and his family, a part in preserving animals alive—giving Noah a role to play as a representative of the human race. Remember, God made man in the image of God and wanted man to rule over the earth and its life forms (Genesis 1:26–28). It is the job of mankind to care for God's creation. And it is for this purpose that God gave mankind, Noah and his family, a part to play and a job to do in caring for the animals during the flood.

And it is God's way to use His servants, Noah in this case, to accomplish His work.

God often uses His servants, even imperfect, limited servants, to do His will and accomplish His work. God sometimes works directly, to do what His servants are not able to do, but He also often works *through* His servants, using them. God uses angels to do His will and deliver His messages. He uses apostles and prophets to deliver His messages and teach the Church, helping to build His Church. Christ builds His Church (Matthew 16:18), but He does some of this building *through* the work of His ministry, imperfect as that ministry might be. Christ baptized, but did it *through* the work of the disciples who did the actual physical labor of baptizing under the authority and direction of Christ (John 4:1–3).

Likewise, God destroyed the surface of the earth in a flood and preserved some animals alive so their species or "kinds" could be preserved and continued. But He did some of that work through Noah and his family.

God never gives His servants a job to do that they are not able to do. It was not Noah's job to bring the flood itself—he could not do that. It was not his job to go to all the parts of the earth to gather Australian animals from Australia, Asian animals from Asia, South American animals from South America, etc. God did not require that of Noah, so God did that part Himself. Or, who knows if God used His righteous angels, under the supervision of Michael or Gabriel perhaps, to help gather the animals from the different parts of the world to Noah and the ark. But that was not Noah's job.

God gave Noah the job of building the ark, gathering the food, bringing the animals into the ark that God gathered to Noah, and taking care of the animals in the ark during the flood. That was quite a full plate of work and responsibility for Noah. If God blessed Noah, perhaps Noah had the financial ability to hire workers from outside his family to do some of this work in addition to his sons. Nevertheless, just planning and supervising this work was a heavy responsibility. Details are not recorded, but it would not be surprising if Noah had to face a series of difficult and frustrating problems and challenges to overcome.

What Noah could do, he did, and what he could not do, God did, by miracles if necessary.

God did not give Noah the job of gathering the animals from all parts of the world. And there is no record of God giving him the job of bringing those animals back to the parts of the world they came from.

Various questions and issues come up about genetic diversity, geographic distribution, etc. in relation to the flood. It can be interesting to speculate about how these issues could work out physically without miracles. But God was able to perform any miracles necessary to accomplish His purpose of restoring the animals on the earth after the flood as they were before the flood. And no miracle God performed was as great as the flood itself or as great as the restoration of the surface of the earth and all life forms during the six days of creation week. If God was able to do that, He was certainly able to solve any problems in repopulating the entire earth as it was before the flood.

There are many solutions to supposed problems with the ark that anyone can search out in their research, but if there is any problem that could not be solved in natural terms, God was able to take care of it with whatever miracles were required. Miracles were required anyway in the gathering of the animals from all over the earth, their peaceful entry into the ark, and the flood itself.

# Carbon-14 Dating

In the last chapter I mentioned radioactive dating that helps establish the age of the earth and the age of various strata and the fossils contained in the strata. For dating rocks in the range of millions of years, this often involves measuring the decay products of radioactive elements that occur naturally in the earth's crust, such as uranium.

Carbon-14 dating is a little different. It cannot be used to measure the age of fossils in millions of years. Its time-range for measurement is much shorter, in the thousands and tens of thousands of years. And it is not based only on the fundamental radioactive elements in the earth's crust.

Anyone can research the details of how carbon-14 dating works, but here is a quick summary.

Cosmic rays, which are particles, bombard the earth's atmosphere from space from all directions. Some of these particles strike carbon atoms in the carbon dioxide in the atmosphere, changing them to isotopes of carbon. These isotopes decay at a certain rate, and when the carbon isotope decays, the carbon is changed back to its original form.

In living plants and animals, there is a constant exchange of carbon with the earth's atmosphere. So the percentage of the carbon that is the isotope formed in the atmosphere from collision with cosmic rays is about the same in plants and animals as it is in the atmosphere.

But this exchange stops when the plant or animal dies and the carbon begins to revert, through radioactive decay, to its original state. Scientists know the rate of decay, and by measuring the percentage of isotopes of carbon in a long dead plant or animal, they can estimate how long it has been since the plant or animal died and no longer exchanged carbon with the atmosphere.

But the accuracy of this depends on the constancy of the cosmic rays that reach the earth's atmosphere. It takes time for the carbon in the atmosphere to stabilize.

If the rate of cosmic ray bombardment of the atmosphere was reduced in the past, that could cause current estimates of the ages of remains of plants and animals to be too old.

And this is possible. Before the six days when God renewed the face of the earth, the earth was in darkness. Something blocked sunlight and a view of the moon and stars from reaching the earth's surface, and whatever blocked sunlight and a view of the sky could have blocked cosmic rays. If this was the state of the earth for any great length of time, that would cause current estimates based on carbon-14 dating to be wrong.

## Summary

The creation account in Genesis can be trusted, literally. There was a world before the six days when God renewed the face of the earth, and that world lasted millions of years. But that natural world, like the world today, was filled with a competitive, cruel ecology, filled with animals killing animals, and the Bible makes clear that this is not the ideal in God's sight. Why?

The physical evidence in fossils, DNA, and existing plants and animals shows intelligent design, but not always righteous, perfect design. And it shows a designer or designers with limited powers who took millions of years and even shortcuts to get results.

Yet God is infinite in intelligence, wisdom, righteousness, and power. Why would He design and make an imperfect creation, or why would he allow it if He did not do it Himself?

That is the subject of the next chapter.

# CHAPTER FIVE

# The Role of Angels in Intelligent Design

## The Problem

How can Genesis be reconciled with the scientific evidence for common descent and emergence of species over millions of years that evolutionists claim point to evolution as the process by which the species came to be?

If you are to deal honestly with the Bible, and if you believe God inspired the Bible as His word and have faith to believe that He will never lie to us but only tell us the truth, you cannot "spiritualize" away the first chapter of Genesis and claim it is only metaphor, that the six days of creation never literally happened.

Genesis, like the rest of the Bible, must be taken literally, except in those cases where there is clear evidence in the Bible itself that it is intended figuratively. There is no evidence, in Genesis or anywhere else in the Bible, that God is speaking figuratively in the account of the six days of creation in the first chapter of Genesis.

Genesis does not say that God created the planet earth in six days. The earth already existed before God began working with the surface of the earth in the six days. And there was apparently some kind of destruction that occurred prior to the six days, a destruction that made the surface of the earth "without form, and void," as the King James Version words it—in chaos and confusion, in other words, in darkness and covered with water.

Lucifer (light bringer), who became Satan (enemy) after he sinned, was given a throne on this earth. Probably, millions of angels were placed on this earth under Lucifer's supervision (Revelation 5:11; 12:3–4).

God must have given those angels some work and activity to do. God would not leave them to be bored. That work could also have been a test of loyalty for Lucifer and his angels to test their character and to test if they would obey God and live righteously.

Lucifer was perfect in his ways till iniquity was found in him. He started out living righteously. But at some point in time he sinned, and he and the angels who followed him in sin rebelled against God (Ezekiel 28:14–18; Isaiah 14:12–15; Revelation 12:3–4).

This may be the root cause of the destruction of the surface of the earth. Lucifer and the angels who followed him in his sin, who became demons, may have continued in sin for a long time before the surface of the earth was destroyed, just as mankind continued in sin for many years before God brought the flood upon the earth in the days of Noah. But when the time came, the surface of the earth was destroyed.

It is not clear if God destroyed the surface of the earth in response to the ongoing sins of Satan and the demons or if Satan and his demons themselves destroyed the surface of the earth in an attempt to destroy all of God's work. The result would be the same.

So there was a time before the six days of creation when the earth existed, and that time that could have lasted hundreds of millions of years. And that time could account for the evidence science claims indicate a very old earth.

But where did the fossils of extinct species such as the dinosaurs come from?

There could have been life on the earth during that time.

If this planet has existed for billions of years as scientists say, then there would be plenty of time to account for the emergence, and in many cases extinction, of thousands or millions of species as shown in the fossil record. And all this could have happened in the period before the destruction of the surface of the earth described in Genesis 1, verse 2.

The fossil record and the record of DNA in existing species, especially non-functional DNA, seems to show the gradual emergence of species through common descent from other species, going back, perhaps, to a single life form, perhaps a single celled organism simpler than any cell existing today. And the period of time that existed before Genesis 1, verse 2 could have been long enough to account for this gradual emergence through common descent.

Would this be evolution?

No, not necessarily.

For this to be evolution, there is another element needed, as I have shown. For this to be evolution, the genetic changes that resulted in new species would have to have arisen entirely through random mutation and spread through natural selection only. There is no room in the theory of evolution as it is believed and taught by the predominant scientific community for any intelligent design in this process or any influence of God or the supernatural.

And there is a problem with that, even apart from the Bible. One might argue that even though the Bible teaches that God is creator, God may have *used* evolution, including random mutation and natural selection, to produce the species. In other words, God could have designed the laws of physics, chemistry, and biology that He has built into the physical world to create an environment on the earth in which evolution would naturally occur. Once He made the universe and this planet with the right laws and conditions, He could just wait, even hundreds of millions of years, for evolution to run its course, and He would not have to intervene through most of that time.

Then, after the destruction of the surface of the earth, in six days God renewed the surface of the earth, preparing it for life, and then restored the same species of life that existed at the end of the fossil record. And it would have to be the same species, at least in general, or there would be a noticeable gap between the most recent fossils (but significantly older than six thousand years even with a wide margin for error, say fossils that are 50,000 years old) and current species. Scientists have found no such gap, as far as I know.

But that explanation won't work, for two reasons.

One, there is the evidence of irreducible complexity in existing species. There are too many features of various species that must work in combination or they have no survival value. All these features could not be produced all at once from random mutations because there are too many of them. They would require an impossible combination of mutations happening together—impossible because the odds against them occurring together are astronomical.

Yet these features could not occur separately because they have no survival value separately. They would individually die out as soon as they start.

Irreducible complexity seems to show that evolution is impossible, even in the time frame of hundreds of millions of years. Gradual emergence of species over hundreds of millions of years through common descent is not necessarily impossible if God used this to make the species. But there would have to be some intelligent design involved—the actual design and changing of DNA from one generation to the next—to account for complex systems with interdependent features that have no value individually.

Could an ordinary insect-eating bird produce the woodpecker as offspring, a bird with many complex interdependent systems able to *work together* to enable the bird to hear ants in a tree, drill into the tree with its beak, then catch the ants on its long, sticky tongue? Yes, but according to critics of evolution, not by random mutation. According to them, the genetic changes would have to be designed to appear in one generation. Intelligent design is possible, even with gradual emergence of species through common descent, but not evolution with random mutation and natural selection only.

One species could give rise to another species if the genetics of the original species were changed. But in many cases, such a change is impossible if not intelligently designed.

I do not say that random mutation and natural selection cannot occur at all. In some cases, a species may change or give rise to a new species entirely through such a mechanism. But not in all cases. There are too many examples of irreducible complexity that random changes cannot account for.

There is another problem in saying that God used evolution, including random mutation and natural selection only, to make the species.

If God did that, He would have designed, from the beginning of the creation of the universe, the laws of physics, chemistry, and biology that would result in all the species of life that has existed—past and present. He would have designed and made those laws for the specific purpose of producing all the species of life through evolution—all the species we see today and have seen in the fossil record.

But those species, many of them, could not be the kind of species God would want.

The Bible teaches that Christ will return to this earth to set up the kingdom of God (Revelation 20:4–6; 3:21). Christ will reign on the earth with the saints for one thousand years. This thousand year reign is called the millennium, and there are many passages in the Bible, especially in the books of the prophets and in the psalms, that describe the peace, happiness, and joy that will exist over all the earth. Satan will be banished—put in a condition of restraint—and will be unable to deceive people or tempt them to sin during this time (Revelation 20:1–3). Christ and the saints will teach mankind the right way to live, and there will be no war during this time (Micah 4:3).

There is one passage I want you to notice. " 'The wolf and the lamb shall feed together, The lion shall eat straw like the ox, And dust shall be the serpent's food. They shall not hurt nor destroy in all My holy mountain,' Says the LORD" (Isaiah 65:25).

This is important. During the millennium, the nature of the animal world will be changed. Predator animals will no longer hurt and kill and eat other animals. There will be peace even in the natural world.

This is important because it shows God's *intent*. It shows what pleases God even in the animal kingdom. Not animals killing animals, but animals getting along peacefully.

If this is God's will, then it raises a question.

Why would God design animals to kill and eat other animals?

Even if God used evolution to produce the species, why would He use that method, with the laws of physics and chemistry He built into the natural universe, to produce the results we see today and the results we see in the fossil record? He could have done things differently. He could have, by whatever method He chose, produced an ecology and a system of plants and animals where all animals live peacefully with each other.

Instead, we see animals killing animals today, and we see the same thing in the fossil record, including giant meat-eating dinosaurs like Tyrannosaurus rex that had existed for millions of years.

Why?

The Bible does not give an explanation to these questions, not directly.

But what explanation is even possible?

Many students who must study evolution in school and who also want to keep faith in the Bible cannot answer these questions entirely to their satisfaction. Many have not thought of and have not been taught even a single *possible* explanation for these things.

But I want to offer an explanation.

## How Intelligent Design May Have Occurred

What I am going to suggest answers the question, why would God use common descent over millions of years to produce the species when He has all knowledge, wisdom, and power and could design and create all the species individually and all at once, in one generation, not over millions of years and not through common descent?

I also will answer the question, why did God make the animal kingdom so competitive and destructive, with predator animals killing other animals, when His description of the millennium in the Bible shows that His will, His preference and what pleases Him, is that animals *not* kill other animals but rather live peacefully with each other just as people should live peacefully with each other?

## Did the Angels Have a Role in Intelligent Design?

I have already shown that Lucifer had a throne on the earth (Isaiah 14:12–15; Luke 4:5–8). God placed him on that throne, probably immediately or soon after the creation of the planet earth hundreds of millions of years ago. If he had a throne, he ruled. Who did he rule over? He likely ruled over a portion of God's angels also placed on the earth, perhaps one-third of all God's angels.

God placed Lucifer and a portion of the angels, perhaps millions of them (Revelation 5:11; 12:3–4), on this earth for a purpose. There was a job for them to do. God uses His angels as servants to participate in God's work, just as God used the angels Michael and Gabriel to deliver messages and perform certain tasks in Bible times. The job that God gave Lucifer and the angels under Lucifer's supervision was probably also a test of their character to see if they would obey and be loyal to God and live righteously. God taught them to live righteously, but God gave them free moral agency and did not force them. They had to make a choice, just as people have to make a choice today.

What was that job?

I suggest that the job God gave the angels was to beautify the earth by developing the variety of species of life that God wanted.

God designed and created the earth and the universe, without physical life, directly. He directly designed the laws of physics and chemistry that would allow physical life to exist. But the earth, while beautiful in its way, was originally without life. It would be beautiful with majestic mountains, clear streams and blue rivers, lakes, and oceans, and beautiful skies, but without the beauty of green grass, trees, flowers, birds, and animals.

God then placed Lucifer and perhaps one-third of all the angels—millions of them—on the earth. Earth would be their home. He gave them a great job to do—useful work—but also interesting and enjoyable work that would keep them occupied and let them use their great minds and the abilities God had built into them. They would work in an organized way, as a team, and Lucifer would be their team leader. God would instruct Lucifer, and Lucifer in turn would instruct the angels on the earth and supervise their work.

Then God made the first life form, a simple cell, perhaps of a simple form of algae. That first cell would contain DNA, which God would have designed, and the protein needed to process that DNA. It would probably be much simpler than any life form or cell in existence today. It would contain in its structure all the necessary parts to be a successful life form and a model of more complex life forms to come.

Then God instructed Lucifer and the angels in how DNA and physical life worked. He gave them the knowledge and the power to work with life, to understand the DNA code and how to apply it, and to modify it to produce new species. And God probably showed them the possibilities—what could be accomplished not only with single-cell organisms but what could be done when cells are combined into larger, more complex organisms.

Then God gave them the job to produce all variety of life for beauty—to beautify and complete the making of the earth with beautiful plants and animals. And the nature of the animal world God intended would be noncompetitive. Plants would provide food for animals, and animals would all be vegetarian. Animals would not hurt and kill other animals. They would be given the instinct or biological limitations to control their own reproduction rate so it would not be necessary for predators to limit the populations.

All this could have started *before* Lucifer sinned and began to lead the angels under his supervision to sin.

This would have been during the time described by Ezekiel when Lucifer was "perfect in your ways" (Ezekiel 28:14–15).

Then Lucifer and the angels began to work with the life God had made.

They began to develop new species by making small modifications to the DNA of certain specimens. Each time they made a modification, they could have spent as much time as they needed to make all the changes to the DNA to produce a new species that was fully functional. In other words, they used their intelligence to *design* the changes to make a new species.

They would have started slow. As brilliant as angels may be, they are not infinite like God. They would have to learn as they went, starting slowly till they gained more experience and expertise.

At first they would work only with single-cell life forms, such as algae, bacteria, and one-celled animals such as the amoeba. Eventually, they would be taught by God or would learn on their own how to make multi-celled organisms. They could start by modifying the DNA of a one-celled organism for the cells to adhere to each other after splitting. Later, they could change the DNA so that the cells that adhered to each other would branch off into different tissue types, and that could be the beginning of very simple multi-celled organisms. .

God could have taught and supervised them in detail as they worked and developed more advanced life forms. When they were ready to develop flowering plants, for example, God could have shown them, or made for them, the first flower and told them to use their creative abilities to design and make a huge variety of flowers for beauty, to help beautify the earth as God intended.

It would have been a test for Lucifer and the angels in many ways. They would be tested to determine if they would be loyal to God and obey Him as He instructed them in the kind of life forms to produce. They would be tested to see if they would cooperate with each other in a spirit of teamwork or if they would quarrel among themselves.

They would have to exercise teamwork to get along harmoniously.

Each new species might require the efforts of a team of several angels working together for a period of time, perhaps years. In producing a new flowering plant, for example, one that would be made for beauty, you might have several members of an angelic team. One might be the designer, who would be gifted by God in artistry, who would design the flower for beauty. Another might be the plant designer who would make sure that the plant would live and grow and reproduce successfully, but without spreading out of control and interfering with other plants. Another might design the actual DNA code sequence to produce the plant as designed. Another might actually make the modifications to the DNA according to the new DNA design. Another might coordinate with other angelic teams to make sure this new species of flower does not upset the existing ecology and interfere too much with other plants. And there would be a leader of the team to make sure everyone on the team worked together in a spirit of cooperation.

God is able to give angels the abilities and talents they need to do the jobs He gives them to do, just as He gave humans the talents, wisdom, skills, and abilities to do the artistic work of building the tabernacle and its furnishings in the wilderness after coming out of Egypt. "See, I have called by name Bezalel the son of Uri, the son of Hur, of the tribe of Judah. And I have filled him with the Spirit of God, in wisdom, in understanding, in knowledge, and in all manner of workmanship, to design artistic works, to work in gold, in silver, in bronze, in cutting jewels for setting, in carving wood, and to work in all manner of workmanship" (Exodus 31:2–5).

Why would they use common descent to produce new species from existing species? Why not design every new species with brand new DNA code from scratch?

Making modifications to the genetic code of some individuals of a parent species to make a new species is far easier than designing all new DNA from scratch. An analogy would be a computer programmer or software engineer making a new computer program or software package that is similar to an existing one, with slight differences. The programmer would not normally write a whole new program from scratch. He or she would take an existing program, change some lines of code, recompile it, and produce the new program. That is far less work than writing a new program from scratch. DNA is very much like computer code.

Is this possible? Could angels accomplish this?

Why not? If God gave them the ability, plus His instruction and guidance, why not?

Angels have minds and powers perhaps many times greater than that of humans (2 Peter 2:11). They do not die, so they can continue to live for thousands and millions of years, always learning, always increasing in skill and knowledge.

Look at what man with his powers small in comparison to angels has been able to do in just a few hundred or a few thousand years with no other tool than selective breeding.

Dogs are all one species. Man has not yet changed one species into another. Yet, the vast variety of breeds of dogs can be an illustration. Using nothing else but selective breeding and having no knowledge of genetics and DNA for most of the last few thousand years, man has been able to develop a huge variety of breeds of dogs of every size, shape, color, level of intelligence, physical ability and stamina, and temperament. Only recently has man begun to understand DNA, and now he is making changes to food plants through genetic engineering and recombinant DNA.

If a few humans, over a few thousand years, can accomplish even that small result, how much more could tens of millions of angels, with minds and powers greatly exceeding that of man, accomplish over hundreds of millions of years with God's help, instruction, and guidance?

Yes, there is every reason to believe that the angels on the earth were able to do this if God gave them the job and the ability they needed to do the job. And if they obeyed God and lived righteously, their work would have been a work of joy with an ever increasing sense of accomplishment and purpose as they worked together in harmony and unity, obeying God, and bringing ever increasing beauty to this earth, which was their home.

## Lucifer and His Angels Did Not Pass the Test

Lucifer and the angels God placed on the earth could have produced species of life and an entire ecology of beauty and happiness as God intended. Animals would not be competitive with other animals, killing and eating each other, sometimes in gruesome and painful ways, but animals would live peacefully with each other, as the Bible says they will in the millennium after Christ returns to rule the earth God's way. Lucifer and the angels themselves would be living lives of joyful happiness under God's care. They would be living God's spiritual way of life—the way of love toward God and love toward each other, their neighbors.

But at some point in this process, perhaps after millions of years, Lucifer sinned. His heart was lifted up and he became consumed with vanity. He became God's enemy, and his mind became twisted and evil.

God must have taught Lucifer the right way of life, and Lucifer started that way, for the Bible says he was perfect in his ways until he sinned.

But God gave Lucifer free moral agency. God did not force him to live righteously. Lucifer was able to choose. And he chose the wrong way. He sinned against God and became Satan the devil. He rebelled against God and became God's enemy. His name was changed to Satan, meaning enemy.

But Satan did not stop with his own sin. He began to persuade the angels under his supervision, perhaps a third of all angels, to join him in his sinful ways, and those who did, perhaps most of them, became demons.

But God did not remove Satan and the demons from the earth. They continued to work with life, making new species by modifying DNA. But now, the species were not entirely according to God's original plan and intent. Animals were designed to kill and eat other animals. It truly became survival of the fittest. In modern metaphor, it became a "dog eat dog" world.

There could be a whole history of events that occurred in this time that are not recorded in the Bible. God may have warned the angels and even brought about mass extinctions from time to time before the final destruction of the surface of the earth.

It may not have been God's original intent that the angels develop the huge dinosaurs, and the extinction of the dinosaurs may have been a correction from God. This kind of thing may have happened many times in many ways over hundreds of millions of years.

But God allowed it for His purpose. He was preparing the earth for mankind, and He was teaching lessons and preparing to teach lessons at the same time, both for angels and for mankind. More about that later.

## The Final Result

When the sins of Satan and his demons reached their peak, when the species of life became as they are today—when God's time had come—things reached a climax. Whether by the direct action of God or by the destructive action of Satan which God allowed, the surface of the earth was destroyed. The earth became covered in water and was in darkness, and all physical life was wiped out.

That is the state of the earth as described in Genesis 1:2.

Then, in six actual 24-hour days, God restored the surface of the earth and brought back the life forms that existed just prior to the destruction.

God did not need to change the DNA code of the species that the angels, now become demons, had designed. He did not have to re-invent all new life forms. The real work of the design of the millions of species is in the DNA code, and that design was complete.

Satan has never been removed from his throne till this day. While the surface of the earth was physically destroyed, Satan and his demons remained. They are still on the earth. Satan was on his throne, ready to tempt Adam and Eve into sin in the garden of Eden (Genesis 3:1–19). He was still on his throne in the days of Jesus Christ (Luke 4:5–8). And he is still on his throne today, deceiving the world (Revelation 12:9), and he will remain on his throne till Christ returns to rule the earth and Satan is put away (Revelation 20:1–3).

I have said before that this is speculation. The Bible does not directly say that Lucifer and his angels were given the job of developing species of life. But the Bible *allows* for this possibility. And scientific evidence seems to indicate that species of life emerged through common descent over millions of years.

In the next chapter, I will discuss in more detail how this possibility can reconcile science with the Bible.

# CHAPTER SIX

# The Advantages of Guided Modification and Branching as an Explanation

## Give It a Name

I call the process I described in the last chapter, "Guided Modification and Branching" (GMB). If you prefer, it can be called, "Designed Modification and Branching," or "Incremental Design through Common Descent," or "Intelligent Design through Common Descent," or "Intelligently Designed Modification," or "Descent with Designed Modification." Its defining feature is the deliberate and intelligent design of the DNA code that makes up each species, through the work of angels under God's authority and supervision, not all at once, and not with each species designed individually from scratch, but over a long period of time with the DNA code from a parent life form modified to produce each new life form.

The process of incremental design from generation to generation may have been suggested by others before, with God directly and alone making the design changes. But I am adding a feature that explains the *reason* why

God used a process that took millions of years, a reason why God used a process that modified DNA from one generation to the next to make each new species or kind of life and a reason why God allowed an ecology and a kingdom of animals hurting, killing, and eating other animals—a system different from the system God will have in the millennium when "they shall not hurt nor destroy in all My holy mountain" (Isaiah 65:25).

That reason is simply this. Rather than design and bring into existence all species of life directly Himself, God chose to use some of His angels as servants and tools to help accomplish this work. And because angels, as powerful as they are, are limited, the work had to be done over a long period of time. Modification of species from one generation to the next is easier than designing the DNA code from scratch for each species. God is infinite and could have done everything quickly Himself, but with the angels, the work had to be done in a way and at a pace they could manage.

It was not just the development of each individual species that had to be managed. An entire planet-wide ecology had to be managed and kept in balance. That is enough of a challenge to keep the angels productively occupied for millions of years. Each species had to be kept in balance with the rest of the ecology.

Did this happen? Were things always kept in balance and in harmony? Not necessarily. At some point, Lucifer rebelled, became Satan the adversary of God, and began to lead his angels into rebellion. This process could have started small with a few leading angels, then began to spread as more and more angels joined Satan in his sinful ways and attitudes. The total corruption that came to all those angels who followed Satan's lead and example could have taken millions of years. But Satan's way is not peaceful. There would have been conflicts and competition even among the angels, even as they began to develop competitive, conflicting species. The design of the natural world mirrored in some ways the corruption of the character of Satan and his angels who became demons.

Jesus said to the Pharisees or Jews of His day who opposed Him, "You are of your father the devil, and the desires of your father you want to do. He was a murderer from the beginning, and does not stand in the truth, because there is no truth in him. When he speaks a lie, he speaks from his own resources, for he is a liar and the father of it" (John 8:44).

God is in absolute control. Satan is not more powerful than God. God put limits on how far Satan and his angels could go in a wrong direction, and the various extinctions we see in the fossil record, including the mass extinction of the dinosaurs, could have been examples of God's corrective restraint on what Satan and the angels were developing on the earth. Scientists surmise that an asteroid strike on the earth caused the extinction of the dinosaurs, and God could very well have caused this event as His corrective judgment on the direction and the extent of the errors Satan and his angels were making in the design of the species. He didn't correct everything wrong they were doing, but He corrected some of the most dangerous developments.

One thing is evident. It would have been difficult, if not impossible, for man to live on the earth and "be fruitful and multiply; fill the earth and subdue it; have dominion over the fish of the sea, over the birds of the air, and over every living thing that moves on the earth" (Genesis 1:28) in the presence of giant meat-eating dinosaurs like Tyrannosaurus rex, popularly known (and loved by children worldwide) as T. rex. With such monsters on the earth, it would have been difficult even for man to stay alive.

It was because of the sins, rebellion, and corruption of Satan and his angels that many of the species they developed were competitive, predatory animals that killed other animals. That is simply a reflection of the corrupted character of Satan and his demons themselves. God allowed it because He saw He could use it for His purposes, purposes that included the making of mankind and the lessons the human race needed to learn in this world, lessons God can use for eternity.

I talk about this more later.

# What Happened During the Six Days?

Some may wonder what actually happened during each of the six days of creation.

While the description of creation week is literal, not figurative, it is told from the perspective of someone on the earth, not someone looking down on the solar system from outer space.

I don't claim to understand everything about this account perfectly. Some details we may not know in this life. As Paul said, "For we know in part and we prophesy in part" (1 Corinthians 13:9). Some questions may have to wait till the return of Christ to be answered, and He can explain all things to us completely and perfectly. Until then, we will always have some questions about the events and certain passages in the Bible. And we must always have a willingness to let God's word teach us. We must *grow* in grace and *knowledge* (2 Peter 3:18) and not have a mind that is closed to God's word and unwilling to learn new things (Hosea 4:6; Matthew 13:52; Luke 5:39), yet have faith to trust God that He is true and faithful and all things have an answer even when we don't yet know what that answer is. Abraham believed God, and it was accounted to him as righteousness (Genesis 15:6; Romans 4:3; Galatians 3:6–7; James 2:23).

For what it is worth, here is what I understand about the six days of creation.

"In the beginning God created the heavens and the earth" (Genesis 1:1).

As I have explained, the original creation of the earth could have occurred hundreds of millions or even several billions of years ago. The creation of the sun, moon, stars, and the entire universe could have occurred at the same time, as far as what the Bible says. The earth was originally in a beautiful state, ready for life to be made even before life existed. God placed Lucifer and perhaps a third of all angels on the earth and gave them a job to do (just as later God gave Adam a job to do—see Genesis 1:28; 2:15). God then started the first life form, instructed the angels, and gave them the job to further beautify the earth by developing, over time, millions of species of plants and animals. God may or may not have revealed to them that they were also preparing the earth for the creation of man. While this was a gift of service to the angels, service and work they would enjoy, it was also a test of their character, a test they could pass or fail. They had free moral agency and had to choose.

"The earth was without form, and void; and darkness was on the face of the deep. And the Spirit of God was hovering over the face of the waters" (Genesis 1:2).

At some point, Lucifer and his angels turned from God's righteous way of life, and they sinned. Lucifer became Satan the devil, and the angels that followed his sinful ways became demons. Eventually, as a result of sin, the surface of the earth was destroyed. All living things died (fossils remained in the ground, however). The earth was covered in water and was in darkness.

The darkness could have been caused by several possible things.

One possibility is that the atmosphere of the earth was so polluted with dust and debris that it was no longer transparent to light. Another is that there was a dust cloud surrounding the earth in space, one that blocked the light from the sun, moon, and stars. There could even have been a dust cloud that engulfed the entire solar system, not only blocking light from reaching the earth from the sun, moon, and stars, but blocking light from the sun from even reaching the moon to be reflected off its surface.

"Then God said, 'Let there be light'; and there was light. And God saw the light, that it was good; and God divided the light from the darkness. God called the light Day, and the darkness He called Night. So the evening and the morning were the first day" (Genesis 1:3–5).

Whatever the cause of the darkness, God cleared enough of the dust or whatever was blocking light from the sun to allow a diffuse kind of light over the earth. The rotation of the earth, whether it had continued through the destruction or had stopped and was started once again, provided the day and night cycle of light and darkness. This was the first day, a literal 24-hour day as we understand days.

"Then God said, 'Let there be a firmament in the midst of the waters, and let it divide the waters from the waters.' Thus God made the firmament, and divided the waters which were under the firmament from the waters which were above the firmament; and it was so. And God called the firmament Heaven. So the evening and the morning were the second day" (Genesis 1:6–8).

The "firmament" can be understood as an expanse. It would be the atmosphere, especially the lower atmosphere. The water above the expanse would refer to water-bearing clouds and water vapor, which would be both in the atmosphere and above much of it, and the water under the expanse would refer to the world-covering ocean at that time. The "heaven" here refers to the atmosphere. In the Bible, there are three heavens (2 Corinthians 12:2–4). There is the atmosphere (Matthew 24:30), the heaven of space that includes the solar system, the stars, and the galaxies (Exodus 32:13), and the heaven of God's throne (Matthew 6:9; 2 Corinthians 12:2–4). This was the second day.

"Then God said, 'Let the waters under the heavens be gathered together into one place, and let the dry land appear'; and it was so. And God called the dry land Earth, and the gathering together of the waters He called Seas. And God saw that it was good" (Genesis 1:9–10).

In the third day, God separated the water from the land. He could have done this by removing some of the water to lower its level or by raising land areas above the surface of the water.

"Then God said, 'Let the earth bring forth grass, the herb that yields seed, and the fruit tree that yields fruit according to its kind, whose seed is in itself, on the earth'; and it was so. And the earth brought forth grass, the herb that yields seed according to its kind, and the tree that yields fruit, whose seed is in itself according to its kind. And God saw that it was good. So the evening and the morning were the third day" (Genesis 1:11–13).

Still in the third day, God restored the plant life that existed before the destruction of the earth. He did not redesign the DNA code. He did not need to "reinvent the wheel" as they say. The plants and their DNA code had been designed as a balanced, working ecology. God did not discard the work of the angels, but used it, restoring plant life as it was before.

The Bible shows that God does not waste things or opportunities. He uses them for good. He uses even mistakes and sins to teach us lessons. When Christ miraculously provided food for the crowds from a few fish and loaves of bread, after the crowds had eaten, He instructed that the fragments be gathered so nothing was wasted (John 6:11–13). King David, a righteous man God called a man after His heart (Acts 13:22), sinned in the matter of Uriah the Hittite. He committed adultery with Uriah's wife Bathsheba, then arranged for Uriah to be murdered to cover his sin when Bathsheba became pregnant. David then took Bathsheba as a wife for himself (2 Samuel 11:1–27). God punished David for his sin with multiple punishments (2 Samuel 12:9–23). Their son died, yet later, God used that marriage to give David another son by Bathsheba, Solomon, who became a great king of Israel and one of the wisest men who ever lived.

Lucifer rebelled against God and became Satan, but God has used Satan's rebellion to accomplish His purposes.

Satan sinned against God, yet God used Satan to teach Job lessons (Job 1:6–22; 2:1–10; 42:1–6).

My point is, God does not waste work accomplished if it can be used for a constructive purpose. He didn't waste David's marriage with Bathsheba even though it came as a result of David's sin. And God did not waste the work Lucifer and his angels accomplished in designing a working ecology. He used the genetic code they had designed to restore the same species of plants, and later animals, that existed just prior to the destruction of the surface of the earth.

"Then God said, 'Let there be lights in the firmament of the heavens to divide the day from the night; and let them be for signs and seasons, and for days and years; and let them be for lights in the firmament of the heavens to give light on the earth'; and it was so. Then God made two great lights: the greater light to rule the day, and the lesser light to rule the night. He made the stars also. God set them in the firmament of the heavens to give light on the earth, and to rule over the day and over the night, and to divide the light from the darkness. And God saw that it was good. So the evening and the morning were the fourth day" (Genesis 1:14–19).

The "making" of the sun, moon, and stars here in the fourth day probably does not refer to their original creation. Remember, this account is explaining the creation from the point of view of someone on the earth. At the beginning of the fourth day, while enough dust was cleared from the atmosphere and maybe also space to allow diffuse light, the sun, moon and stars were still not visible. God completely cleared the dust so that the sun, the moon, and the stars could be seen clearly. And if, in the destruction, the orbits and positions of the earth around the sun and the moon around the earth had been disrupted, God restored the positions and movements of the earth and moon to their proper place. God did not "make" the sun and moon in the sense of bringing them into existence in this fourth day, but He *made them to be lights for signs and seasons.* He did this by restoring the bodies of the solar system to their proper orbits if necessary and by clearing any dust that blocked a clear view of them.

"Then God said, 'Let the waters abound with an abundance of living creatures, and let birds fly above the earth across the face of the firmament of the heavens.' So God created great sea creatures and every living thing that moves, with which the waters abounded, according to their kind, and every winged bird according to its kind. And God saw that it was good. And God blessed them, saying, 'Be fruitful and multiply, and fill the waters in the seas, and let birds multiply on the earth.' So the evening and the morning were the fifth day" (Genesis 1:20–23).

In the fifth day God restored the species of birds and fish and other water creatures. Just as with the plants, He restored the same species with the same genetic code that existed before. He did not change the DNA that had been developed, but brought those species back to life as they were. He did not waste the work the angels had performed, even thought those angels turned to sin.

And there would be no long time gap in the fossil record between the animals that existed before the destruction and the animals that exist today. To scientists who look at fossils and DNA, it would look continuous, as if there was no destruction and the plants and animals continued uninterrupted for hundreds of millions of years until now.

"Then God said, 'Let the earth bring forth the living creature according to its kind: cattle and creeping thing and beast of the earth, each according to its kind'; and it was so. And God made the beast of the earth according to its kind, cattle according to its kind, and everything that creeps on the earth according to its kind. And God saw that it was good" (Genesis 1:24–25).

In the sixth day, God restored the air-breathing land animals. Again, He did not redesign the DNA code, casting away all He had accomplished through the work of the angels on the earth. He used that work. He used the design embedded in the DNA code.

Even though that design of animals included predators that killed and ate other animals, God saw how he could use even that for a good purpose, to teach mankind lessons, lessons that would pay off for eternity. I will talk more about this later.

And then God made man.

"Then God said, 'Let Us make man in Our image, according to Our likeness; let them have dominion over the fish of the sea, over the birds of the air, and over the cattle, over all the earth and over every creeping thing that creeps on the earth.' So God created man in His own image; in the image of God He created him; male and female He created them. Then God blessed them, and God said to them, 'Be fruitful and multiply; fill the earth and subdue it; have dominion over the fish of the sea, over the birds of the air, and over every living thing that moves on the earth'" (Genesis 1:26–28).

Did man exist before the destruction of the surface of the earth? No. Man was a new creation.

What about fossils and even stone tools and cave paintings that scientists have found that suggest some human-like creatures existed tens of thousands of years ago? Were they human beings like us that existed before the destruction of the earth as stated in Genesis 1, verse 2?

No. Assuming they are correctly dated older than six thousand years, those creatures no doubt did exist before the destruction, but they were not human, though they may have appeared to be like us physically. They would have a certain level of animal intelligence, perhaps far greater than even chimpanzees—enough intelligence to build tools and make cave paintings. But there is one ingredient they did not have that made all the difference. They did not have a human spirit in them.

The Bible teaches that there is a spirit in man, a spirit that can be called the human spirit. It empowers the human mind to know and understand the things humans can know but animals cannot. "But there is a spirit in man, And the breath of the Almighty gives him understanding" (Job 32:8). "For what man knows the things of a man except the spirit of the man which is in him? Even so no one knows the things of God except the Spirit of God" (1 Corinthians 2:11).

That spirit God has placed in man not only empowers the mind of man but makes it possible for him to have a spiritual relationship with God that animals cannot have.

Adam was the first real human being. Eve was the second. The human spirit placed in them made them different from all animals that came before, no matter how similar some animals may have been to humans in body structure.

And it is likely that God not only gave Adam and Eve the human spirit, but also that God directly modified their DNA from the prior human-like creatures that existed before the destruction to give their physical brains added intelligence and ability.

Notice how God worded the making of Adam compared to the restoration of animal life. With the animals, God said, "let the earth bring forth," but with man, He said, "let us make man in our image." Man was specially made by God. This was on the sixth day.

"And God said, 'See, I have given you every herb that yields seed which is on the face of all the earth, and every tree whose fruit yields seed; to you it shall be for food. Also, to every beast of the earth, to every bird of the air, and to everything that creeps on the earth, in which there is life, I have given every green herb for food'; and it was so. Then God saw everything that He had made, and indeed it was very good. So the evening and the morning were the sixth day" (Genesis 1:29–31).

Notice that God says He gave plant life to the animals to eat. I had said before that I thought the animals that were restored were restored with exact same DNA code and design as the animals that existed just before the destruction, including predator animals that killed and ate other animals. But that could be wrong. Animals that we now know as predators, such as lions and wolves, could have had their DNA and design changed some-what so that they were peaceful and only ate plants. This verse suggests the possibility that the animals that existed before Adam sinned and before the flood may have been peaceful. As I said, there are things we do not know for sure and will find out when Christ returns. But either as they were restored to the earth on day six, or later after Adam and Eve sinned by taking the forbidden fruit, or even after the flood in Noah's day, animals such as lions and wolves killed and ate other animals just as Satan and his demons designed them before the destruction.

Also, notice that God saw everything He made, and it was very good. Someone might ask, if God kept the DNA code and the design of animals that included animals killing animals, contrary to God's will for the millennium when there will be peace all over the earth, how could God see it as "very good?"

The answer is, it was very good for God's purpose. Satan was on a throne, and God chose not to remove him from that throne even after he sinned and even after the destruction of the surface of the earth and then the restoration of life and the making of the human race. Why? I explain this in more detail later, but it is God's purpose to teach mankind certain lessons, including the lesson that Satan's way of vanity, hostile competition, selfishness, hatred, and disobedience toward God does not work. It produces misery, destruction, and death. The human race has yet to fully learn that lesson, but it will. And God has left this world under Satan's influence and leadership for the purpose of teaching that lesson.

A word about the making of the woman. God took a rib from Adam and made it into a woman to be Adam's wife (Genesis 2:21–22). What is the significance of that?

It does not mean that Eve was only made from the material in the rib. She would have been a very small woman in that case. But God took a part of Adam's body for the DNA. I suppose any part would serve that purpose because the DNA code is in all the cells. God may have used a rib for the lesson of the rib being close to the man's heart. In any case, the rib had all the DNA God needed, with some modification, to make a woman nearly identical to the man.

The genetic difference between a man and a woman is that a man has an X chromosome and a Y chromosome, while the woman has two X chromosomes and no Y chromosome. God could simply use the X chromosome from Adam, double it, and leave out the Y chromosome to make the basic genetic pattern for the woman. The added material could have been from the earth, but her genetic design was made from Adam's rib using his DNA. Because a man has both an X and Y chromosome and a woman has only two X chromosomes, you can make a woman from a man but not a man from a woman. God may have no doubt made other changes in the DNA to provide genetic variety for their offspring, but the basic DNA was the same.

## Possible Proof Text in the Bible to Prove Life Existed on Earth before the Six Days

Here is something many Bible students may not have noticed before.

Look again at Genesis 1:24. Notice the way it is worded. God made the animals, each "according to its kind." What does, "according to its kind," mean?

Often, when we see the phrase, "according to its kind" or "after its kind," in the Bible in reference to plants and animals, we may think of reproduction. Each kind of plant or animal reproduces according to its kind. This is how it is used in Genesis 1:11–12 regarding reproduction of plants. The offspring of a plant or animal is the same as its parents. This phrase indicates that a pattern is followed—what results is the same as what precedes. Cows produce cows, dogs produce dogs, chickens produce chickens. It's a backwards reference to the parents of offspring.

But Genesis 1:24 is not talking about reproduction. It is talking about creation. Each animal was created "according to its kind." Does this not suggest that a prior pattern, a pattern that defines the "kind," was being followed in the creation?

What does it mean to *make* something after its kind?

It means that it is made according to a previously known pattern.

If you are making something new that never existed before, you are not making it after or according to any previous kind. You are making it new.

But if you are making it after a kind, you are making it after a previously existing and known pattern.

How was each animal created? According to a pattern. What pattern? The pattern of its kind. Each animal was created according to its kind, a kind that must have existed first, prior to the six days. Otherwise, why say it was made according to its kind? For if the creation of animals in the six days was brand new with no precedent, why say animals were created according to any pattern? Why a backwards reference to what was before as implied by the phrase, "according to its kind" or "after its kind?"

Not every student of the Bible will understand this the same way. I leave it to the reader to judge.


## Questions Answered and Problems Solved

This explanation I offer, which I call Guided Modification and Branching (GMB), answers several questions and problems that arise in trying to reconcile physical evidence with the Genesis account of creation.

Should the creation account in Genesis be taken literally? Yes. Though the description of events during the six days are from the point of view of an observer on earth, the six days are six, literal, 24-hour days.

How can anyone who takes the Genesis account literally explain fossil and DNA evidence that indicates species of life arose through common descent from common ancestors over hundreds of millions of years?

There was a period of time, which could have been several billions of years as far as the Bible account is concerned, after God created the planet earth and the universe in Genesis 1, verse 1 and before the destruction of the surface of the earth as described in Genesis 1, verse 2. During that time, species of life could have emerged over hundreds of millions of years through common descent from parent species, with genetic changes being made from generation to generation to produce new species, just as the evidence shows.

If random mutation and natural selection are the mechanisms for one species giving rise to a new species as science claims, why are there not always transitional forms in the fossil record that show one species gradually giving rise to a new species? And why are there cases of irreducible complexity with many characteristics that must work together before any one of them becomes an advantage for the species?

The answer to both of these questions is the same. Random mutation and natural selection are not the main mechanisms for change. Intelligent design is.

Atheistic scientists do not want to accept that, but there is no other way to explain the evidence.

When a species produces a new species through genetic change, a whole collection of changes, made to work together to produce an advantage for the new species, are intelligently and deliberately designed and implemented in a single generation, or a few generations. That is how irreducible complexity comes to be, and that is why transitional forms are sometimes missing. While the whole family tree of species emerges gradually in the larger scale of hundreds of millions of years, many individual species emerge abruptly, in a single or a few generations, from a parent species with no transitional fossils being found.

GMB explains missing transitional forms better than random mutation and natural selection. It also explains the relatively sudden appearance of new forms, such as in the Cambrian explosion, better than the theory of evolution.


If God is designer and creator of all the species of life, and if He is infinitely wise and powerful, why would He take hundreds of millions of years to produce the natural world, and why would He use common descent to produce new species when He could easily design and make each species individually and all at one time?

God is certainly capable of doing just that, but God made the angels for a purpose, as servants to do God's will and help accomplish His purpose. It is God who creates, but He can use others as tools for His purposes.

And while God is infinite in power, the angels are not. They had to proceed more slowly at a pace and with methods they could handle according to their ability, and this was God's will for the angels. He gave them the job of developing the species of life on the earth under His supervision and direction, teaching them and giving them the abilities they needed to do the job. This would actually be a gift to the angels, for it would give them exciting and interesting work to do that would challenge their abilities and give them a sense of accomplishment, and it would give them an opportunity to express themselves creatively.

So the angels designed changes to the species to produce new species a few species at a time over millions of years. New species were not designed from scratch but were modified from existing species, because that is easier, just as a computer programmer often uses an existing program and modifies it rather than write a whole new program from scratch. Shortcuts may have been taken to save time, and the angels could have taken the easiest way with the fewest changes to make new kinds of life and new species. This can be an explanation for vestigial organs, such as leg bones in whales, if it is simply easier to make changes that diminish the feature rather than eliminate it entirely. We do not know the details.

I mentioned the analogy of computer code and how a computer programmer might modify a program to make a new program. If he wanted to get rid of a particular function, he might simply add lines of code to bypass coding for a certain function and not take the time and trouble to search out and "clean up" all the code for that function and remove it.

If God created the species, and if He is all good, why did He design and create predator animals that hurt and kill and eat other animals?

God did create the species, but He did it through angels. They had free moral agency, and they could obey God and work in harmony with His righteous way of life, or they could sin and rebel against Him. And at some point, under Satan's leadership, they sinned and turned against God. And as part of their rebellion, and as a reflection of their own competitive and destructive natures, they designed and produced predator animals that hurt and kill. That was not God's original intent, but He allowed it in the past and in the present. But when Christ returns and Satan is put away (Revelation 20:1–3), the nature of animals will be changed and they will no longer hurt and kill each other (Isaiah 11:6–9).

The job given to the angels on the earth, besides giving them enjoyable work and activity to do in beautifying the earth, which was their home, also served a purpose to test their character to know if they would remain righteous or turn to sin. God could instruct them and encourage them in righteousness, but they had free moral agency, and they had to choose for themselves. Lucifer chose to sin and rebel against God, and he became Satan, the enemy of God (Isaiah 14:12–15; Ezekiel 28:16–18). He also led many or most of the angels on the earth to follow him in his sinful rebellion, and they became demons (Revelation 12:3–4).

So they failed the test.

It is interesting to speculate about what might have happened if Lucifer and the angels on the earth had remained righteous, if they beautified the earth and produced exactly the kind of animals God wanted, animals that were peaceful with other animals, and if Lucifer and the angels on the earth lived righteously as God instructed them, loving God and loving each other. In other words, if they passed the test, even after hundreds of millions of years, what would have been the result?

This earth is only one planet in a solar system of several planets, and our sun is only one of perhaps ten thousand billion billion suns in the universe. There could be billions and billions of planets that can be made to support life. Yet there is no evidence science has found so far to indicate that physical biological life exists anywhere in the universe but on planet earth.

Is it possible that, if the angels had passed the test of character, God might have used them, with the knowledge and experience they had, to bring life to the countless other planets in the universe? The Bible doesn't say, but that is an interesting possibility to consider. But they sinned and failed the test.


There seem to be parallel animals of certain characteristics from different lines of descent. Marsupial mammals exist predominantly in Australia and carry their young in pouches, while placental mammals exist primarily in the rest of the world. These appear to come from different lines of common descent. Yet, there are parallel kinds of animals in each group, such as placental squirrels and marsupial squirrel-like creatures. How can this be accounted for?

With GMB, this can easily be accounted for. Placental and marsupial animals are different branches—different lines of descent. Placental mammals have a common ancestor different from the common ancestor of marsupial mammals.

But with intelligent design by teams of angels over millions of years, information and ideas could be exchanged. A team that is developing placental squirrels could exchange information with a team that is developing marsupial squirrel-like creatures in Australia. This information exchange could include body structure, behavior instincts, body chemistry, and the DNA code needed to implement the design. And this information exchange could occur by teams of angels voluntarily sharing information to help other teams or simply by teams observing what other teams were doing.

During the time angels were developing species, Satan rebelled against God. His rebellion spread to other angels on the earth as he persuaded them to follow his ways. But this could have taken millions of years, and during that vast time period, some angels may have followed Satan relatively quickly while other angels remained righteous for a time or followed his rebellion in varying degrees. So at any time, you could have a mixture of demons fully committed to Satan's way and angels who still tried to be righteous. So in the development of species, you could, at any one time, have a mixture of competition and cooperation among the angels. But probably, eventually all or most of the angels on the earth joined in Satan's way and became demons.

Is there a way to test Guided Modification and Branching scientifically? Does this idea make any predictions?

Maybe.

The evidence I have offered so far should demonstrate the case. Irreducible complexity and the sudden appearance of new forms without transitions requires intelligent design, but the evidence of a savage competitive world, extinctions, and vestigial organs shows that the design is not perfect, and common descent shown by DNA and fossils and the age of the fossil record shows that this design did not happen quickly. This points to an imperfect, intelligent designer or designers, and the angels on the earth, acting as God's agents but limited in ability and flawed in character, could produce such a design.

But there may be other evidence.

It may be possible to find evidence of changes for a designed purpose in the direction of that purpose before that purpose is achieved. I don't know how hard it would be to find such evidence. I suppose the only real history of changes is in the fossil record, and that limits us to structural changes only, not chemistry.

But suppose you find a change in the structure of a species, a change that has no apparent advantage for an organism. But it is a stepping stone to something else. Then later, more change occurs and a definite advantage to the species appears. The first change had a purpose, but that purpose did not become evident until many generations later. It was a change in a *design direction* for a future benefit, but that benefit was not realized till many generations and many other changes later.

If this happens quickly, we will not see evidence, for there will be no transitional forms. But if the time period between the first change and the future changes that give purpose to the first change is long, we may see evidence of this in transitional fossils.

Perhaps some have found such evidence already. I do not know.

You can also have cases where species are not entirely competitive, but cooperative. You have examples of symbiosis, where animals cooperate to their mutual benefit, or even to the benefit of one species but not the other.

An example of symbiosis that many people have observed in nature videos is the example of certain birds picking the teeth of crocodiles. The crocodile lays there with its mouth open and the bird goes *inside the croc's mouth* to pick food from the crocodile's teeth. This benefits both species. The bird finds a source of food and the crocodile gets its teeth cleaned to prevent tooth decay.

But perhaps here may be another case of irreducible complexity. At some point the birds would have to develop the instinct to go inside a predator's mouth for food, while the crocodile would have to develop the instinct not to eat the bird in one quick bite. How could this have developed through random mutation and natural selection? Even if a bird had a mutation that caused it to go inside the crocodile's mouth, unless the crocodile had a mutation to not eat the bird, *at the same time,* that particular bird would become extinct in a second.

But there can be other cases of cooperation besides symbiosis.

Some species may limit their reproduction to stay balanced with the ecology. Instead of competing with all other species and perhaps overrunning their environment when there is an absence or shortage of predators to limit their growth, their reproduction rate goes down. It is as if the species voluntarily limits its reproduction so as not to overrun or overcrowd or create problems for other species. I do not know how random mutation and natural selection—survival of the fittest—can explain that, but GMB explains it easily.

# Objections

Some might say, "By crediting the creation of species to angels, you are denying God's role as creator." No, I am not. God is still the creator, but He has designed and created the species *through* the angels to whom He gave that job. They were His agents, doing the job He gave them to do, not perfectly, because they sinned, but overall, yes.

There is a passage in the Bible that illustrates the principle that one can do something *through* and *by* someone else. Christ baptized more disciples than John, yet Christ Himself did not do the physical work of baptizing, but His disciples. "Therefore, when the Lord knew that the Pharisees had heard that Jesus made and baptized more disciples than John (though Jesus Himself did not baptize, but His disciples), He left Judea and departed again to Galilee" (John 4:1–2).

The above scripture does not contradict itself. Christ did not directly do the physical work of baptizing. His disciples did the baptizing. Yet, that work is attributed to Christ when the passage says that Christ baptized more disciples than John. Jesus did not do the physical work of baptizing, but He baptized in the sense that His disciples baptized in His name and by His authority. Christ builds His church (Matthew 16:18), yet He uses the work of the apostles and entire ministry to help Him build the church (Luke 5:10; John 21:15–17; 1 Corinthians 3:5–16; Ephesians 4:11–16).

Who judges the world, Christ or the Father? One scripture says that the Father judges no one but has committed all judgment to the Son. "For the Father judges no one, but has committed all judgment to the Son, that all should honor the Son just as they honor the Father" (John 5:22–23). But another scripture seems to say God the Father judges us. "And if you call on the Father, who without partiality judges according to each one's work, conduct yourselves throughout the time of your stay here in fear" (1 Peter 1:17). The answer is that God the Father judges the world *through* and *by* Jesus Christ. Christ does the actual judging, but the Father judges in the sense that Christ judges by the authority of the Father and according to the Father's command. The fact that the Father judges *through* and *by* Jesus Christ is stated more directly in Acts 17: "Truly, these times of ignorance God overlooked, but now commands all men everywhere to repent, because He has appointed a day on which He will judge the world in righteousness by the Man whom He has ordained. He has given assurance of this to all by raising Him from the dead" (Acts 17:30–31).

Likewise, when God gave the angels the job of developing the wide variety of species of life on the earth, the angels did the actual work of developing the species, but it was done by the authority and in the name of God. By giving the job to the angels and by giving them the ability and the instruction they needed to do the job, God was developing and creating the design of all the species *through* and *by* the angels on the earth. That is the principle of delegation.

And even with the work of the angels, God probably directly created the first one-celled life form with DNA and proteins that the angels first worked with to start the process of making new species of life. The angels probably did not create the first life, but God delegated to them the job of modifying what He had created to produce a huge variety of life forms.

Besides that, God directly populated the earth with the plants and animals we see today during the six days of the creation week, but He did it according to the DNA design work the angels had developed. He did not waste their designs but used them. The design work was already complete.

Technically speaking, the species that exist today are not actually descended from a common ancestor, but are descended from the plants and animals God directly restored during the six days of creation week. But they appear to be descended from a common ancestor because these plants and animals are based on exactly the same design as the plants and animals that were descended from a common ancestor before the destruction of the surface of the earth described in Genesis 1, verse 2. So they match up with the same DNA and they match up with the fossil record.

Moreover, there was a brand new creation during the six days of creation—man. Human beings did not exist before God made Adam. There were animals before the destruction of the earth that appeared similar to man and had similar DNA, but they were not the same. God could have modified the DNA, especially that part having to do with the brain and with intelligence, and God added an ingredient that was new—the human spirit in man (Job 32:8; 1 Corinthians 2:11). That spirit imparted a power of intellect that animals do not have. It gave man the ability to have a relationship with God on a spiritual level when combined with the Holy Spirit God gives as a gift to those who obey Him (John 14:16–17; Acts 5:32; 2 Timothy 1:7). That human spirit may also impart human consciousness and free will to the physical human brain.

The first man, Adam, was a unique creation that God made on the sixth day of creation week.

God is creator.


Someone might say, "By claiming that the design for the species of life came from common descent over millions of years as evolutionists claim, you are advocating the theory of evolution."

No, I am not.

The theory of evolution, as taught by majority science, is more than common descent over millions of years. It is the theory that this common descent occurred due to natural, random forces only and never through intelligent design. Darwinism is the theory that the development of species occurred due to random variation and natural selection, not design by any intelligent being or supernatural entity.

What I am suggesting is a form of intelligent design, not evolution.

Someone might say, "There are examples of evolution that seems to have occurred through random mutation and natural selection only, and there are no missing transitional forms or irreducible complexity added to a new species in those cases."

That may be true. Even many intelligent design advocates acknowledge what may be called "microevolution" in some cases. An animal species changing color to match its environment might be an example of this. A modern example might be the resistance some bacteria develop to antibiotics, or the resistance some insects develop to pesticides (like bed-bugs).

So, random mutation and natural selection sometimes play a part in the development of the characteristics and features of species. Probably natural selection's main function, besides minor adaptations to a changing environment, is to eliminate error and repair faults to the DNA code that harms the species. But it does not account for the development of all species with their many cases of irreducible complexity and missing transitional forms. Only intelligent design can account for that.

Random mutation and natural selection can and do play a limited role, but they are not the main mechanism for change in a species or kind that results in a new species or kind. The main functions of natural selection are to weed out harmful errors in the DNA code and provide for minor adaptations to a changing environment.

# Conclusion

The suggestion and theme of this book, that God used angels to help design the species through intelligently designed changes to DNA to make new species from parent species over millions of years, is speculation. The Bible does not say this directly. It is an attempt to reconcile what the Bible does say with the physical evidence science has found. I believe it succeeds in that purpose. It enables a literal understanding of the Bible to be reconciled with the physical evidence science has found for the emergence of species through common descent over millions of years.

I cannot say this is the only possible explanation. But it is the only one I know of that works. If my speculation is wrong and if there is some other logical way to reconcile the Bible with scientific evidence, I do not know what it is. I have not found any other explanation that makes sense.

Why even speculate about this at all? Why not just say there are some things we do not understand, and we trust that God's word is true? Some may have the faith to do just that, trusting God that there is an explanation they do not know and cannot imagine. But that does not work for everyone. God gave us minds that can think and reason and work out problems. God gave us a natural curiosity about things. We naturally strive to find answers to questions that puzzle us, and that is normal and good. And for some, having just one possible explanation in mind for reconciling science and the creation account in the Bible is helpful for strengthening faith in God's word. It is better than no explanation at all.

If students in the science classrooms who want to believe the Bible cannot reconcile the Bible with the physical evidence they learn about in class, that can injure their faith in God's word.

But this is an explanation that makes sense.

That is why I wrote this book.

There is a cultural war being waged in the United States. There is a great anti-God movement that manifests itself in various ways. It is part of a liberal, leftist agenda that has been gaining ground tremendously in the decades since World War II.

The theory of evolution is part of that battleground. The dominant secular education system, the scientific community, and liberal media in this country have been pushing the theory of evolution on our children and young people in the public schools and colleges and on the country as a whole.

## Why God Allows Suffering

Several times in this book I mentioned that God has allowed Satan's evil influence and the suffering in this world to continue for the purpose of teaching mankind lessons. I think at this point I should explain that in more detail.

God's purpose for mankind is to give us eternal life in His kingdom. There are implications, though no definite statement in the Bible, that our ultimate destiny may be to work with the planets in the universe, imparting and developing life on them according to God's plan and direction, as Lucifer and the angels failed to properly do on the earth.

But God does not want another rebellion in His kingdom, not ever. Lucifer at some point evidently decided that the way of vanity, conceit, pride, selfishness, and rebellion against God and His authority was a better way of life, a life that would make him happier than God's way of love toward God and neighbor. God does not want His human children to make the same mistake. So He has made a plan to teach us lessons.

God has appointed Satan to rule over mankind for six thousand years since Adam sinned. During this time, God allows Satan to deceive and lead mankind into living a sinful way of life, a way of vanity, pride, hostile competition, greed, selfishness, hatred, lying, murder, rebellion and disobedience toward God, and every kind of sin. The result has been suffering, destruction, death, injustice, and all the evil we see in the world today. God is allowing mankind to see and experience the results of that way of life.

But at the end of the six thousand years, Christ will return to rule the earth and put Satan away, in a condition of restraint, where he will no longer be able to deceive, tempt, or lead mankind into sin. Christ and the resurrected saints will rule the earth and teach mankind

God's way of love, peace, truth, and obedience toward God. The result will be peace, justice, prosperity, and happiness for one thousand years (Revelation 20:1–5).

People who live during the thousand year reign of Christ will be able to compare the history of human life for six thousand years living Satan's way with the happiness of living the way of God under Christ's rule. They will see that God's way is better.

After this, the vast majority of mankind that had lived and died during the six thousand years will be resurrected back to physical life (Revelation 20:11–13, Ezekiel 37:1–14). They too will be able to compare the happiness of God's way with the suffering of Satan's way.

And all who repent of their sins and accept Christ will be converted and saved, and they will be given eternal life in God's family.

This is why God is allowing mankind to get a bellyful of Satan's way of life—to learn a lesson that will pay off for eternity.

In the next chapter I will suggest practical strategies those who believe in God can use to protect their freedom of religion for themselves and their children.

# CHAPTER SEVEN

# The Battleground

## The Battle between the Theory of Evolution and Faith in the Bible

In the United States, the theory of evolution is strongly promoted and pushed by the scientific community, by the public education system from the grade school level to the most advanced college and university courses, by liberal politicians and judges, and by the dominant news and entertainment media.

This promotion is part of an anti-God, atheistic, and materialistic agenda. Hard-core evolutionists passionately argue their case that, as they say, "Evolution is a fact!" No it is not—these evolutionists speak as if they do not even know what a scientific fact is. They are simply affirming their personal belief and faith that evolution is true.

How are Christians or anyone who believes in God and the Bible and does not accept evolution to cope with this? How are believers in God to respond?

In this country, our Constitution and our legal traditions protect citizens' rights to practice their religious beliefs without government interference. This country was founded in large part by people coming from other countries seeking religious freedom.

But those rights are being eroded.

Government, tax-supported schools teach our children the theory of evolution contrary to what their parents and churches teach them from the Bible.

Students of all ages are confused, many of them. Many of them believe or want to believe in the Bible (though that number is diminishing as religion declines in the United States). But they are faced with challenges in the classroom that they are ill-equipped to face. They cannot answer some of the things their teachers tell them, and they do not know how to reconcile the teachings they receive in the classroom with what they read in the Bible.

They learn about physical evidence of fossils and DNA that their teachers tell them points to evolution, and they cannot explain that physical evidence in light of the Bible account.

This book is intended to fill a gap in that regard and to show how the Bible can be reconciled with scientific evidence.

The rest of this chapter is intended to give some suggestions on how this situation can be understood and handled by those who want to defend their religious faith and their rights to practice their religious beliefs without government interference.

## Answering in the Classroom

Students can sometimes feel intimidated by their instructors of evolution in the classroom. Here are a few points those students should keep in mind.

Science, as a community, is handicapped by the scientific method in determining what is true. The scientific method, *as practiced* (whether or not stated in any written description of the scientific method), does not allow or consider supernatural explanations for known processes or observations. In other words, science, as practiced, *assumes*, without proof or evidence, that there is no God who intervenes in natural processes. It *limits itself* to considering natural causes only.

Science does not consider this to be a limitation because it assumes there is no God who acts in the physical universe. But this is an assumption only, with no proof or evidence to back it up. And refusing, as a matter of policy, to consider God's intervention as a cause is indeed a limitation.

It means science is biased in its interpretation of evidence. It does not look at all possible explanations. Regarding the emergence of species, science closes its eyes to possible evidence that God or any supernatural entities have intervened to produce the species. This is a serious handicap that can result in error, and science's conclusions cannot be trusted.

If science excludes the supernatural, then *it can't know* if life evolved. Why? It can't *consider if it was created!* To know that life evolved, science would have to show that the evidence is compatible with evolution *but not creation*. If the evidence is compatible with both, then the issue is undecided. But science can't consider creation because it says, "it's religion, not science."

You cannot prove something or find the truth by looking at only one side of an issue.

The scientific method, with its assumption that there is no God who intervenes in physical events, does not work well when investigating the origins of things, such as the origin of the universe, the origin of life on the earth, the origin of all the species, or the origin of the human race.

The scientific method works fine in the laboratory. It works fine when studying repeatable processes. That is because God does not often intervene in natural law. God made the physical laws to allow man to live in a predictable environment, one he can work with and control to a certain extent. It is also not God's will *at this time* to force people to acknowledge His existence (this will change). That does not suit His purpose right now, so He stays in the background enough to give human beings the option to reject Him.

Yet, God has left evidence in the creation that He exists. "For the wrath of God is revealed from heaven against all ungodliness and unrighteousness of men, who suppress the truth in unrighteousness, because what may be known of God is manifest in them, for God has shown it to them. For since the creation of the world His invisible attributes are clearly seen, being understood by the things that are made, even His eternal power and Godhead, so that they are without excuse" (Romans 1:18–20).

But science cannot look at the evidence of God's existence as long as it refuses to consider His existence and intervention as a matter of policy.

What is the evidence God exists? Creation itself is the evidence.

Even if evolution were true (and it is not), it does not explain how the universe got here with all its precisely tuned laws and constants that permit the existence of life and ourselves. It also does not explain human consciousness, which is a total mystery to science.

Creation is proof of God (Psalm 19:1–4; Romans 1:18–20), and the consciousness of our minds is part of that creation, and science has no answer for it.

And when discussing consciousness, scientists rarely discuss the problem intelligently. They confuse consciousness with intelligence or self-awareness or attention. They will scan a subject's brain while that subject thinks about something, and they will note what areas of the brain "light up" on their scanners indicating that portion of the brain is most active. But that says nothing about consciousness, what produces it, and why it exists.

Only God can create consciousness, and only God can make a universe perfectly tuned and suited to providing an environment in which man can exist and live.

Some atheists may suggest that the reason this universe exists and has laws that permit man to exist is that all possible universes exist and the only one we know of is one in which we are able to exist. The idea of the multiverse is attractive to atheistic scientists as a way of denying God's existence, and it may be a motivation for the intense interest many scientists have in string theory, because string theory allows for multiple universes.

But this is a false argument. If atheistic scientists thought all possible universes exist, there would have to be universes in which God exists, and therefore God may exist in this universe. But atheists do not consider that.

The scientific community is dominated by atheists, people who by choice and by their faith reject the existence of God. They are sure God does not exist because that is what they choose to believe, nothing more.

Sometimes, in discussing this, atheists will argue against God's existence by saying, "The suffering and injustice in this world proves that God does not exist, for if God exists and is good and all-powerful, He would intervene to stop injustice and suffering."

But there is an explanation for this, and the Bible helps us understand it, as I have pointed out in the last chapter. God is allowing the suffering to help the human race to learn lessons, lessons that mankind has not yet fully learned but will learn, and those lessons will pay off in happiness for eternity.

I think the real reason many atheists choose to believe there is no God is because they do not want to face the truth that there is a God who has authority over their lives and will judge them for what they do. They want their freedom to live their lives as they want without being haunted by a feeling of guilt or fear. They prefer the illusion over the reality.

An atheist might ask me, "Why does God exist?"

I can ask, "Why does anything exist? Why is there not eternal nothingness?" But it is evident something exists.

When I meet God after Christ returns and sets up His kingdom or I meet God in the judgment, I may ask Him why He exists. I might also ask Him about the eternal past, something I cannot comprehend. There are some things I cannot understand now with the limitations of my human mind.

Another thing students need to keep in mind is that they are not obligated to give alternative explanations to their teachers for the evidence their teachers claim proves evolution.

Scientists and science teachers in the public school system are trained and indoctrinated in evolutionary theory. They are trained to look at all evidence through the lens of evolution. They have spend years developing explanations for and interpretations of the evidence that fit the theory of evolution. And then, a teacher might say to a student, "How can you explain this piece of evidence without evolution?" and demand that the student instantly give an alternative explanation in a few minutes or even a few days.

But for a student to really evaluate the evidence fairly, he or she would need to see the *original data,* all of it, not data limited and filtered by advocates of evolution, and then have years, as the scientists have had, to interpret and understand what the data really shows.

If you are a student and a science teacher asks you a question about God or about how to interpret evidence the teacher says points to evolution and you do not have an immediate answer, don't let that intimidate you. You are not obligated to answer every question immediately.

That does not prove evolution.

And if you are asked to believe in evolution, you have a right to require proof. Science may claim, "Science doesn't deal in 'proof,'" but that is a dodge. Evolution is not just a private issue for the science community that chooses not to deal in "proofs." Evolution is a science issue, a religious issue, and a legal issue. Courts of law deal in "proof," and people have a right to try to prove for themselves the things they believe that affect their religious traditions and their moral views.

Some colleges give courses in formal logic.

With formal logic, you can sometimes prove conclusions if you first set the premises. Geometry and other specialties in mathematics deal in "proof." Can science prove evolution according to the rules of formal logic or mathematics? The only way they could "prove" evolution that way is if they first set as a premise the assumption that there is no God who acts in the universe.

There is no way you can prove the nonintervention of God in the natural world.

And remember that evidence never points to anything unless it is first interpreted. And if the person or community that interprets the evidence is biased, as science is, their interpretation should be suspect.

## The Schools, the Communities, and the Courts

There is a legal battleground and there is a battleground in your mind and the minds of everyone, but especially students.

In the legal arena, communities through their state legislatures and probably in some cases individual public schools and universities have tried to limit the teaching of the theory of evolution or mandate the teaching of intelligent design. In some cases, attempts have been made to prohibit the teaching of the theory of evolution. In other cases, attempts have been made to give equal treatment to intelligent design, letting students hear both sides and make up their own minds.

Both of these kinds of cases have been blocked and overturned by liberal courts, in my opinion, wrongly. The courts, for whatever their reasons, have in effect sided with the forces of anti-God liberalism that is in the process of turning the country from Christianity to atheism, and materialistic, secular evolution is one of their tools. Court justices are appointed long-term, and this trend of the liberal courts is not likely to change in the short term.

Courts are wrong to exclude intelligent design studies from public school science classes. They seem to have made their decision based on the motivations of those who advocate intelligent design. Motivation has nothing to do with it. But if they are going to look at motivations, why not look at the motivations of the atheistic scientists who want to exclude ID for the purpose of promoting their atheistic faith?

There is an opportunity lost here, for science and its companion discipline mathematics can contribute much through probability theory to determine if evolution is even possible in face of the evidence.

In truth, it is the teaching of evolution as "fact" in the public schools that is unconstitutional, for it seeks to impose atheism on public school students. The government and government controlled and supported schools should be neutral in regards to religion, neither standing for religion or against it. But evolution, taught as dogma and absolute truth, denies God's existence, because evolution can only be known to be true if there is no God. If there is a possibility God exists, there is a possibility He used intelligent design to produce the species. And if that is possible, then it is possible that the theory of evolution is false. Therefore, to say that evolution is certainly true is the same as saying there is no God. And this has become government policy because of liberal, anti-God courts.

I wish not to get into giving advice on how to fight this politically. I will leave that to others. Instead, I will concentrate on how students, and others, can protect their own minds. As public policy, it is a political fight, but it is also a fight for the hearts and minds of people one person at a time. There may be political considerations to what I say, but I am focusing on each individual.

Know your rights. Be informed and aware. Be clear in your own mind what is right, what is true. Study the issues as necessary until you are clear about them. No one has the right to mandate how you think.

Whatever the liberal politicians try to do to advance the atheist agenda and trample on the people's rights, it becomes easier when people are not well informed or are unclear in their thinking.

Know what your rights are, and more importantly, in cases where there may be ambiguity, know what your rights *should be* in a free society. Make up you mind if you believe in freedom, the freedom Americans have fought for in history, and decide to what extent you believe in freedom. Stand for your freedom to think and draw your own conclusions.

People's freedom is certainly threatened. We are letting the liberals erode it, little by little. Our ancestors crossed an ocean and tamed a wilderness for their freedom, and as a nation we have fought wars to protect it. But we are losing it because we tolerate losing it. We are losing it because the liberals have a stronger desire, a stronger will, more energy, more zeal, to take it from us than a conservative majority has to keep it.

Those who are aware of what is happening know that this is not just about the theory of evolution. But this book is about the theory of evolution, which is one weapon the atheists use.

And the liberals and atheists can be ruthless in their efforts to force their anti-God faith and their agenda upon you, through laws, through court rulings (even unconstitutional laws and court rulings), and through various pressure tactics. Scientists and science teachers can lose jobs and career opportunities for their beliefs.

No matter how much control anti-God liberals attain in government, education, and the media, they cannot control how you think and what you believe, even though they want to. That is your primary defense, I think.

## Students' Freedom of Thought and Belief

One of the things that is a constitutional right or *should be* a right of all students in public schools is a right to believe what they choose.

This can simply be called, freedom of thought, or freedom of belief. If not termed that in the Constitution and our legal traditions, it can certainly be justified by rights that are defined: freedom of religion, freedom of expression, freedom of speech, and freedom of the press. These are all manifestations and expressions of our basic freedom to think, to make our own decisions, and to choose our own faith.

Freedom of thought is the basis for all these other freedoms.

The government is not your mommy and daddy. They do not have the right to control you, to tell you how to live your life, except to maintain public order. They do not have the right, or the wisdom, to tell you what to believe in.

Many atheists probably think they know better than you how you should think, believe, and live. But the truth is, you know better than they do. They have proven they do not have the wisdom to live their own lives properly, much less tell others how to live.

Some of the most ignorant, foolish people in the world are also the most highly educated in this world's education. Much of their education is nonsense, but they don't know it.

In the public schools, students should not be required to agree with class material. They should be required to be familiar with it. They should know, and be tested on, what the course teaches. But they should not be required to state that they agree with what is taught.

So in the case of the theory of evolution, students should learn and be tested on what the theory says. They should learn and know the arguments evolutionists use. But they should not be required to make statements in test answers or homework assignments that indicate they agree with those arguments.

They can be required to know and say, in effect, "Evolutionists claim that random mutation and natural selection account for the origin of species." But they should not be required to say, "Random mutation and natural selection account for the origin of species."

Right now, students may feel pressured to give answers to test questions about evolution they do not really believe, or to turn in essays or homework assignments they do not really agree with. They are faced with an unpleasant choice: lie in their school work and test answers and say they agree with evolution when they don't, or tell the truth about what they believe, that evolution is false, and risk lower grades or perhaps failure in the course.

The key question is, do schools have the right to mandate students' beliefs?

It's not a question of knowledge and familiarity with the course materials. Students should do their study work and know what the courses they take teach. But they should not be required to agree with the teachings. Private and religious schools are not supported by tax dollars, and no one is required to attend them—they can do what they want. But public, tax-supported schools should not feel they have the right to exercise thought-control over their students.

This is most important regarding the teaching of evolution, but it can apply to anything—economics, business, political science, sociology, history, medicine, physics—whatever. It can and should apply to anything in which a student may have a different opinion on a subject than what is taught in a course.

Why should public, tax-supported, government-controlled schools mandate the personal views, opinions, and beliefs of students regarding the subjects they study and learn about?

And in science, it is not even healthy to build an atmosphere, culture, and tradition where students are pressured to agree with their teachers in everything. Many advances in science have occurred precisely because students have disagreed with the teachings and traditions passed to them.

Our society has made progress in advancing knowledge because we have, till recently, been willing to question and challenge orthodoxy. That is how we learn to go beyond past limitations and past mistakes. That is true in science as it is in other endeavors.

Students should be required to *know* what is taught in the courses they take, but not required to express *agreement* with what is taught in them.

Freedom of thought in the public classroom, freedom of students to agree or disagree with course material and not be penalized for believing what they choose, may or may not be established and maintained by the schools themselves or by the institutions of government. But individual students can be aware of their rights to think as they choose and maintain them as well as they are able. They can be aware that this country was founded on such rights.

In no subject is this more important that the theory of evolution.

# The Declaration of Independence

If you ask someone in the United States, "What is the foundational law in this country?" they would probably answer, "The Constitution." That may be right, in a sense, but I would suggest a different answer.

It is the Declaration of Independence that establishes the United States as a nation separate from Great Britain and no longer under English law. The Declaration of Independence came first and has priority over the Constitution. The Constitution could not exist without the independence of our country that the Declaration helps to establish. The sole purpose of the Constitution is to serve the nation and the principles established in the Declaration. Without our independence as a nation, the Constitution means nothing. Moreover, the Declaration shows the mindset of the same founding fathers that made the Constitution, and it can help clarify the intended meaning of the Constitution.

In my layman's opinion (I am not a lawyer), there may be grounds for considering the Declaration of Independence as a source for legal authority, even equal to or greater than the authority of the Constitution. The Constitution helps implement the Declaration, not the other way around.

And it is useful to keep in mind that our founding legal documents, the Declaration and the Constitution, are not documents written by lawyers, for lawyers.

These documents are for the people and are intended to be understood by everyone.

What does the Declaration of Independence say about God's existence?

"We hold these truths to be self-evident, that all men are created equal, that they are endowed by their Creator with certain unalienable Rights, that among these are Life, Liberty and the pursuit of Happiness."

The Declaration of Independence, a possible source of legal authority in the United States because of our history and its ability to help show the intent of the Constitution due to its proximity in time to the Constitution, states that it is self-evident that God exists and that He is the creator of mankind.

The Declaration of Independence does not try to prove God's existence. I have tried to prove it in this book, but to our founding fathers, it was so self-evident that no proof was required. In other words, it was obvious to them that God exists and that God made the human race.

They had a lot more common sense than many people today.

God says the same thing in the Bible. His existence is obvious. "For the wrath of God is revealed from heaven against all ungodliness and unrighteousness of men, who suppress the truth in unrighteousness, because what may be known of God is manifest in them, for God has shown it to them. For since the creation of the world His invisible attributes are clearly seen, being understood by the things that are made, even His eternal power and Godhead, so that they are without excuse" (Romans 1:18–20).

God says that atheists are without excuse. And He also calls them fools. "The fool has said in his heart, 'There is no God.' They are corrupt, They have done abominable works, There is none who does good" (Psalm 14:1).

No wonder the authors of the Declaration of Independence said that God's existence is self-evident. Creation is the proof of God—the universe with its fine-tuning both for our existence and our observation and appreciation of the universe, our conscious minds, and the irreducible complexity of the living world, including the DNA code that makes proteins and the protein structure that must be already in place to interpret the DNA code to make proteins. All this, of course, was not known to our founding fathers, nor to the Bible writers, but they saw all the universe around them, and they knew there must be a God to create it all.

I don't say that agnostics and young people with open minds seeking answers are foolish. A person with an open mind, willing to learn, is not a fool. It can take time, maybe many years, to think things through, to find answers, including the answer to the question, "Is there a God?" But hard-core atheists who are sure there is no God are just wrong.

The Constitution with our legal traditions guarantees freedom of religion, freedom of speech, and freedom of the press. People under the Constitution are free to believe what they want about God, to believe in Him and worship Him as they choose, or to believe there is no God.

But before that, the Declaration of Independence states that it is *self-evident* that God exists and is the creator of mankind.

In other words, while the Constitution gives citizens the right to not believe in God, the Declaration establishes, *as a matter of law* in the United States, that, in truth, God exists and that man did not evolve.

If the Declaration of Independence has legal authority in this question, what does that mean in practical terms?

It does not mean that religion is mandated or that people should not have the freedom to be atheists. But at the least it should mean that government should not base its decisions on the opposite idea—that there is no God. No government agency or branch or court should base its decisions on atheism—on the belief that there is no God.

The Declaration does not override the constitutional right of freedom of religion. Government cannot require you to believe, or not believe, in God.

But the Declaration of Independence at least establishes a principle that no law or court decision can be based *on an assumption of atheism*. Courts cannot assume there is no God and then base their decisions on that assumption. If they do, they have exceeded their

CHAPTER SEVEN—THE BATTLEGROUND    167

authority, and their decisions are invalid and may later be overturned by a higher court or the same court.

If a court makes a mistake, if a later court realizes the mistake, it should admit the mistake and correct it. It should not be so attached to precedent that it will knowingly allow a mistake to continue.

Notice that the Declaration of Independence says that God exists and the human race was *created*—it did not evolve by mindless forces in a godless universe as militant atheistic evolutionists insist.

While courts are forbidden to favor or endorse atheism, that is exactly what the materialistic science community does. They base their decisions on the assumption of atheism. Only by making that assumption can they assert that *evolution definitely happened through natural causes only, and is truth.* Only by assuming there is no God who is able to intervene in natural processes can they reject intelligent design as "not science." For science cannot "prove" (or "show" or "demonstrate" or whatever word you prefer if you want to say science doesn't "prove" things) that God or His servants did not supernaturally intervene and intelligently design some or all of the changes in life forms from one generation to the next.

# The Science Community

Science, by which I mean the majority science community, has a problem. It is biased against belief in God. This bias prevents it from honestly and accurately seeking and learning truth.

Science practices the scientific method. It does not perfectly agree and articulate what that method is, and all scientists do not strictly follow the method, but it exists. The scientific method exists and is a standard by which scientists are expected, by the majority in their community, to do their work of teaching and researching. And whether stated in any description or not, an important principle in that method as practiced is that scientists are not to consider any supernatural cause for the things they investigate and observe.

Science limits itself to the study of natural processes only.

In some ways, this is a good thing, because science is not equipped to study the things of God.

The problem is, science in its pride and conceit does not regard this as a real limitation. It thinks it is no limitation because it thinks there is no God and no supernatural.

Science has become atheistic. That is the truth. Not all scientists are atheists. Many scientists believe in God, and many of these believe in creation and intelligent design. They are marginalized. They are persecuted. The scientific community—majority science you can call it— tries to exclude scientists who acknowledge God and His intervention in the world.

Hard-core evolutionists often say something like this: "No serious scientist doubts evolution."

That is nonsense. The only way that can be true from the evolutionist's point of view is if he defines a "serious scientist" as one who believes in evolution. That way, if you don't believe in evolution, you are not a "serious scientist." But that is a pretty ridiculous definition of a serious scientist. You can define a scientist in different ways, but I think if a person has a PhD in science at an accredited college or university, has published papers in accepted scientific journals, and is actively employed as a researcher or teacher in science, it is fair to say that person is a scientist.

And there are many with such qualifications who do not agree with the theory of evolution. I do not mean they just disagree with the details of evolution. They do not believe evolution actually occurred. They believe in intelligent design.

But is such a scientist "serious?" I guess the evolutionist would say, if he doesn't accept evolution, he is not "serious." But that is just silly. These scientists are just as serious about what they believe and the scientific work they do as evolutionists.

Don't buy it if an evolutionist tells you all serious scientists accept evolution. That is simply not true.

Individual scientists may believe in God. I think Darwin himself had religious beliefs. But majority science dominated by atheists today does its best to silence and marginalize them.

Majority science is atheistic in its teaching and work.

Representatives of majority science will deny that the scientific community is atheistic. If you ask a scientist, "What does science say about God's existence?," a typical response might be, "Science does not address that question. Science does not say that God does or does not exist." But that is false.

It is a public relations ploy. Such scientists know that many people believe in God, and they hesitate to offend them in such an obvious way as to say that science says there is no God. But that is exactly what majority science teaches. It teaches that there is no God, definitely.

How does science teach that there is no God if it does not openly say there is no God?

It does so by teaching the theory of evolution as definitely true, or as "fact" as some of them might say.

Can the theory of evolution be true if God exists? Certainly. If you don't look at the evidence too closely, you might conclude that God exists and He *might have* let natural forces only produce life forms through evolution.

But science says it is certain that evolution occurred, and part of the theory of evolution as majority science teaches it is that evolution occurred through natural forces only—no intelligent design by God or any supernatural agency.

What science is saying is that there is no chance that God ever intervened in the development of life. That is

the same thing as saying that evolution definitely occurred.

You see, if science was neutral about God—maybe He
exists or maybe He doesn't—it could not be sure about
evolution. It might say evolution is possible, but not definite. That is because, if there is any possibility that God
exists, He might have intervened in the development of
species, directly or indirectly. If He did, then life forms
did not come only by natural forces. But "natural forces
only" is a part of the theory of evolution as it is taught by
science.

So if there is a possibility of God's existence and intervention in the development of life, then there is a possibility that evolution is false (even if it were possible,
which it is not). If science were really neutral about
God's existence, it would acknowledge that the theory of
evolution might be false. Even if it insists evolution is
possible, it would have to acknowledge the possibility
that it never happened.

To acknowledge the possibility of a God who can intervene in physical processes is to acknowledge that evolution might not have occurred. To acknowledge the
possibility of God, one must also acknowledge the possibility of intelligent design.

But when science absolutely rejects intelligent design
and insists that evolution occurred through natural
forces only, which it does, it is showing its true colors. It
is saying there definitely is no God.

Why? Because science has no way of knowing that
evolution occurred if there is any possibility that God

exists. Logically, the only way science can be sure that evolution happened is if they are sure there is no God who could have intervened to intelligently design the species (directly or through His servants the angels).

How can science be sure that species came through natural forces only if they acknowledge even the possibility of God? They can't. Because, if God exists, the species might not have come from natural forces only.

For the sake of public relations, science will not directly and openly say that there is no God. But by saying that evolution through natural forces definitely occurred, they are *indirectly* saying that there is no God. Because, if God *might* exist, then evolution *might* not have happened. If God might exist, then intelligent design might be real. That is logic science won't openly admit.

Atheistic scientists accuse ID advocates of trying to put religion into science. But it is the atheists who put atheism into science.

Majority science bases its conclusions on the assumption of atheism. Science takes definite positions about evolution that would make no logical sense if there was any chance that God exists.

Evolutionists have said, or at least one has said, that we must not be fooled by the appearance of design in the natural world, because, though biology *appears* to be designed, it is not designed. Evolutionists also maintain, in court cases, that intelligent design is an attempt to bring religion and God into science but that religion must be excluded from science.

But if biology appears to be designed, why is it not possible that it was in fact designed? How can science or evolutionists *know* that life was not designed? The only possible designer anyone seriously has suggested is God (and I am suggesting that God did the some of the designing indirectly through the work of angels). How can science know that God did not design the first life form and then supernaturally design and guide the development of species over time?

The only way science can know this is if it knows there is no God or if it knows that God chose not to intervene in natural processes in the past. But science cannot prove that and won't even discuss it.

If science therefore claims evolution as truth or "fact" but cannot prove intelligent design as false, then science is claiming that there is no God who has intervened in this universe in the past—in other words, no creator. Otherwise, science could not say that evolution is definitely true.

Therefore, in claiming that evolution is definitely true, science is claiming there is no creator God. And this is what science—the scientific community—claims.

Atheists therefore are using science to impose their faith in atheism upon students and the public, and they are using tax dollars and public institutions to do it, contrary to the laws that guarantee religious freedom.

Science is biased against God and the supernatural, and that bias diminishes its ability to find truth. Its scientific method does not allow it to examine both sides of the issue objectively. That hurts science and makes it unreliable. It diminishes the credibility of science, and rightly so. It makes it impossible for science to prove evolution even if science did deal in "proof."

The bias of science against God's intervention in the history of the origin of things diminishes science's credibility.

When a scientist says the supernatural is outside the bounds of science, that is sometimes code language atheists use to say there is no God—the supernatural doesn't exist. The limit of not considering supernatural causes is a limitation and a *handicap* of the scientific method, perhaps a necessary handicap, but a handicap and a limitation nevertheless. But atheistic science does not acknowledge that it is a limitation and a handicap and insists that science is the best way of knowing things and that totally naturalistic evolution (no God) is truth and fact. In saying this, science reveals its atheistic foundation.

If the scientific method, as it is practiced by the majority science community, requires that evidence be interpreted on the assumption that a God who can intervene in physical processes does not exist, then the scientific method needs to be changed.

The unimpeded and objective search for truth is more important than the traditions and methods of science.

There may be many reasons why many scientists believe in evolution, but I will mention two that come to mind.

One, the theory of evolution is attractive because its explains the physical evidence—fossils, DNA, vestigial organs, etc.—better than the only idea of intelligent design these scientists are aware of—sudden, perfect creation six thousand years ago, which is the version of intelligent design believed and promoted by many in traditional religion.

GMB is a version of intelligent design that can explain all the evidence, but most people are not aware of it. It actually shares some conclusions with the theory of evolution, but it is not evolution.

Two, some scientists are militant atheists. They have faith that there is no God, and they choose this faith for themselves because they do not want to face the possibility that there is a God who has authority over their lives and will judge them for the choices they make. The theory of evolution is one expression of the atheist's faith, and many of these atheists want to validate their faith by imposing it on others, by force if necessary. These atheists care less about the evidence than about their belief in "no God."

There is reform needed in science. It is out of control. It has allowed itself to become the tool of militant atheists intent on forcing their atheistic faith on society at large. The scientific community is proving to society as a whole that it is unwilling or incapable of policing itself, of controlling itself, or of maintaining high ethical standards and objectivity in the search for truth. It is begging to be regulated.

Freedom is good, but with freedom comes responsibility to use that freedom wisely, and freedom abused often becomes freedom lost.

If science is willing to reform itself, it needs to develop a new scientific method. I am not qualified to articulate all the details of such a new method, but one change is needed. It needs to continue to limit itself to the study of natural processes only—it is not qualified to research the supernatural—but it needs to acknowledge that as a real limitation. It can set a boundary between the natural and the supernatural, and not go beyond that boundary in its work, but it must not base its conclusions on a belief that there is nothing beyond that boundary.

It should accept the study of intelligent design to the extent of determining, without bias, if intelligent design occurred, but it should not cross the boundary into theology by trying to study the nature of a supernatural designer.

It is one thing to decline to investigate non-natural (supernatural) causes. It is something else to assume or declare that there *are no* supernatural causes, which is what science does when it says that evolution is fact (or true) or even when science says that science is the best way of knowing things.

If a theory explains the evidence—if the theory *fits* the evidence—that does not prove or show that the theory is true. You cannot honestly say that the theory is true or "fact" on that basis alone. You must also show that the evidence is *not* explained by the competing theory. If two opposite theories both explain the evidence, then the evidence "points to" neither.

The theory that competes with evolution as an explanation for life and the species of life is intelligent design.

Which theory explains the evidence better, evolution or intelligent design? To answer that question, you have to examine both theories as explanations for the evidence. It is not sufficient to say that the evidence supports evolution. It must also *not* support intelligent design.

If intelligent design explains the evidence equally well as or better than evolution, then you cannot honestly say that the evidence shows evolution to be true.

The fact that militant evolutionism and the mainstream science community resist intelligent design studies and the teaching of intelligent design in the classroom as a possible alternative to evolution shows their hypocrisy.

The modern ID movement does not talk about God or the Bible. It focuses on the evidence from a strictly scientific point of view. It looks at the details of molecular biology and the complex interaction of DNA, RNA, and protein in the cell and the whole organism, as well as larger organs and systems, and it examines this complexity with the tools of mathematics, probability, statistics, and information theory—the same tools used in computer science, cryptology, and other scientific specialties—to see if the un-intelligent processes of the theory of evolution are plausible explanations for the origin of life and species.

Science does not have to get into religion to examine the evidence this way.

The fact that militant evolutionists are so opposed to ID as to invite the courts into the science classroom shows evolution for what it is—what many have been saying—a *faith*, not science, like a kind of anti-God religion. The very opposition of mainstream science to ID shows that the faith of atheism—faith that there is no God who has intervened in physical processes—has captured and taken control of mainstream science to use science as a tool for its ends. And what are those ends? The stamping out or the diminishing of religion.

Science limits itself to the study of the natural world only. It does not study the supernatural.

Science has a right to say it will not study certain things. But it does not have the right to say that the things it chooses not to study do not exist.

Does science do this? Does it say there is no God?

Not openly. Not directly. But indirectly, in an underhanded way, that is exactly what the predominant scientific community says. Science says there is no God. And it uses public funds and government backing to teach atheism to students in public schools.

How does it do this?

By teaching evolution as absolute truth, not to be questioned, or as some wrongly put it, as "fact." That is the same as saying there is no God.

Why?

If God exists, alternative explanations to evolution can exist for life, explanations that involve God's direct intervention or the intervention of His created spirit beings. Darwinism may not be true in this case. There can be supernatural explanations that fit the evidence better than Darwinism. But by refusing to consider that, yet affirming that evolution must be true, science in effect says the supernatural does not exist, for if it exists it might explain the evidence better than evolution. And if the supernatural does not exist, God does not exist.

I know I have been repetitive here, but this needs emphasis.

To say that the theory of evolution, including variation through natural causes only, is definitely true, is exactly equivalent to saying that there is no God. When the majority science community says evolution is definitely true, it is teaching and promoting atheism.

There should be a difference in thinking about repeatable, everyday processes and the origin of things. God does not often intervene in everyday processes. He wants physical processes to be predictable according to laws man can discover so man can control his environment and understand the normal workings of the universe. God maintains the physical laws of the universe and their operation as He designed them because He wants the human race to be able to work with the environment, to work with matter and energy, technology and invention. He wants man to live in a predictable environment so he can work with his environment.

But with the origin of things, it is different. God has intervened in the origin of things, or nothing would exist. Science needs to acknowledge the possibility that God exists and has intervened in the past—to make the universe, to make life and the vast variety of life, and to make mankind.

The scientific method is a disaster and an embarrassment in investigation of origins of things, and more and more scientists are understanding this.

Science, if it is to be reformed, needs a scientific method that studies natural processes only but does not deny the possibility of supernatural forces also.

Science needs a new scientific method that maintains a separation from religion without denying the validity of religion. It needs to refrain from claiming or endorsing supernatural causes without denying of possibility of supernatural causes.

And it must not base its conclusions on the assumption of atheism.

Why does science exclude consideration of supernatural causes? There can be two reasons: because science is not qualified and equipped to study theology and the supernatural, or because the supernatural does not exist. It should be the former, not the latter.

Science is right to limit its study to natural causes only. But it should acknowledge this limitation as a real limitation and not say that it is no limitation because the supernatural doesn't exist. Nor should it base its conclusions on the assumption that there is no supernatural, especially its conclusions about the origin of things.

In the search for truth, there is a need for some of the tools and processes of science and many of the physical things science has discovered, but without the restriction that science imposes that supernatural causes not be considered. Theology and religion do not have the tools of science. Science refuses to look at the supernatural possibility, even to acknowledge that it is possible that there is a supernatural. And science does not consider the Bible to be a credible source of evidence, even though the Bible is God speaking, the same God who *made* the universe and is an eye-witness to everything that has happened.

## Is Intelligent Design Science?

The intelligent design movement (ID) says that there are characteristics of life that are best explained by intelligent design, not random mutation and natural selection. They claim intelligent design is science and should be studied and taught as science.

The atheistic majority in the science community has opposed intelligent design, claiming it should not be considered science and should be excluded from the science classroom. They claim that intelligent design is an attempt to bring religion into science. The liberal court system has sided against intelligent design, ruling against it being taught in the public schools. The court has said that ID is religion disguised as science.

Their reason, part of it, is that part of the motivation of intelligent design advocates is religious.

But that is a ridiculous reason to exclude intelligent design from science. Whether a subject or point of view is science or not has nothing to do with the motivations of the scientists who study the subject or the point of view. In the long history of science, many scientists have had religious beliefs, and those beliefs motivated and informed their studies.

ID advocates and teachers have taken great care to keep religion *out* of ID. ID does not name the designer. It just examines if design is a better explanation for life and the variety of life than undesigned variation and natural selection. The designer could be any intelligent entity, even intelligent aliens from space.

It is atheistic science that has tried to put religion into ID so they can exclude it. The atheists say ID is religion because, directly or indirectly, it points to God and the supernatural.

But if science is to be unbiased in the matter of God's existence, and if it is to be honest in its studies, there is nothing wrong with studies in science that point to a boundary between the physical and the supernatural. That is what science *should* do. Science should identify those cases of historical origins of things—whether life, or the species of life, or the universe itself—that it cannot explain. It should not get into religious and theological details, but it should not insist that the supernatural does not exist.

This is why I say that a new scientific method is needed, one that continues to restrict itself to the study of natural processes but does not claim that the natural is all that exists.

Intelligent design studies natural processes and shows the limitation of unintelligent natural processes as an explanation for the origin of the species of life. But it does not discuss God and the supernatural. It identifies a boundary and acknowledges that there may be something beyond that boundary, but it does not cross the boundary.

That is exactly what a new scientific method should allow. It should find the boundaries between what is possible physically and what may require the intervention of God and the supernatural world, and it should not deny the existence of the supernatural when it finds those boundaries. It should not get into religious and theological arguments and discussions, but neither should it deny that the supernatural may exist.

I just mentioned that atheists try to put religion into ID. But they also bring religious anti-God arguments into their defense of the theory of evolution. They can't do this in the science classroom of course, but they do this in their public discussions, including discussions that take place in public lectures, on the Internet, and in books. Recorded lectures, the Internet, and published books are becoming the new forum of discussion of scientific matters that is replacing, to a degree, published papers in scientific journals (think about that when you are blogging). In the future, I think some scientific thought will be led by books and Internet discussions as much as by scientific papers.

Atheists do this by pointing to the evils of the natural world and claiming that this shows there is no God, for if God exists, He would not allow it. And since there is no God, evolution must be true—there is no alternative. So to defend evolution and atheistic science, they have to cross the border into theological discussions and say, because the world is evil, there can be no God.

So to refute intelligent design, atheistic scientists enter the world of theology and ask, "How could a good and infinite God permit an evil world?" Is that question not outside of science?

Evolutionists exclude God when it suits them but talk about Him negatively (why would He make such an imperfect world?) when it suits them. The fact is, theological puzzles—why would God allow evil?—may play a major role in atheists' and evolutionists' beliefs whether they talk about it or not.

They also focus on six-thousand year old earth creationism as a straw man to argue against, as if that has anything to do with intelligent design. They know they can prove the earth is old, but they cannot explain irreducible complex systems in the natural world.

There are questions relating to intelligent design that must be part of science because only science has the tools needed to investigate these questions and help find the answers. Who should study and do research to determine if intelligent design is true? Religion? Theology? They do not have the tools of mathematics, information theory, biology and genetics, statistics and probability theory, paleontology, geology, cryptology and code-breaking, etc. that science has and may be needed to properly study this question.

Is intelligent design true? Who is qualified to study and answer that question? That question should not be lost in the discussion of whether intelligent design is science.

Is it true? How can we find out?

Science has the tools to help find the answer. But the scientific community, dominated by atheists and hindered by its flawed scientific method, refuses to honestly study the question, probably because it is against the answer it anticipates it may find. Does it refuse to study intelligent design because it suspects that if it does it will discover that it is true, and if intelligent design is true, that is one more piece of evidence to prove that God is real, and atheistic science doesn't want the public to know that? By refusing to study intelligent design, is science trying to hide the truth from the public?

It makes no more sense to separate science from religion than to separate science from mathematics or science from history. If you pursue the truth, you have to allow overlap between studies.

Science assumes, on faith apart from evidence, that intelligent design is wrong. And it tries its best to exclude it from science and to prevent a rigorous study of it.

At its heart, ID is simply the scientific exploration of the question, "Is evolution, the development of life without intelligent design, possible?" It is just as scientific to explore reasons why evolution is impossible as to explore reasons why it is possible.

Science has no right to exclude from consideration a possible explanation for life (God), then insist that such exclusion is part of the best way of knowing things, or to insist that evolution is "fact" (true) because there is no other explanation, when a possible explanation exists (God) but is excluded by science.

Any approach to finding truth that says that there is no God is certainly not the best way of knowing things (Psalm 14:1; Jeremiah 8:9).

Intelligent design *is science,* and that will be more and more obvious in spite of past foolish and shallow court decisions. ID overlaps with too many fields that are already recognized as science or accepted by science, such as mathematics, statistics, probability theory, information theory, computer science, forensics, cryptology, military intelligence, and the search for extraterrestrial intelligent life.

Investigation to determine if something is the result of intelligent design is definitely a part of science despite attempts by atheists to exclude it. It is as much science to examine the chemistry of life to look for signs of intelligence as it is to examine possible signals from space to look for signs of intelligence.

Over much of science's history, scientists accepted God's existence and did not try to use science to promote atheism. Over the last several decades, science has allowed itself to be hijacked by atheists and used for their purposes—to advance their agenda.

But the pendulum may be swinging the other way, now.

Some opponents of ID say ID is not science because it is not a complete theory. But that doesn't mean it is not science. Einstein worked for many years on general relativity before it was complete, but it was certainly science while he was working on it.

The general public is learning to distrust science because of science's bias against God. Science is losing its credibility because of its bias and hypocrisy.

## What Is the Best Way of Knowing—Or, Who Will You Have Faith In?

Science exercises faith. By faith, I mean a chosen belief system.

Faith does not have to be a religious faith. It can be a secular, materialistic faith. One can have faith that there is no God just as surely as one can have faith that there is a God. It is a matter of choice.

Not everything we believe is a matter of choice.

If I walk outside, and I can see that the sun is shining, I have no choice about believing that the sun is shining. I have to believe that the sun is shining. I would have a hard time trying to convince myself that the sky is completely overcast with clouds when the sun is burning my eyes and everything around the sun in the sky is blue. That is not faith.

But if someone tells me that the sun has existed for five billion years, I can choose to believe that or not. If someone shows me evidence, I can interpret that evidence as I choose. I can choose to believe certain things. What I choose to believe can be my belief system, and I can have faith in the things I believe.

Faith can also be trust and confidence in a source of information, trust that the information from that source is accurate. For example, someone can have faith that God exists, and they can have faith that what God says in the Bible is true. God can be their trusted authority and source of knowledge. Belief in God and His word the Bible is one kind of faith.

One can also have faith that God does not exist. One can have faith in science. Atheism is a faith in the sense that it is a chosen belief system. I would not call it a religion, but it is similar to a religion in the sense that it is faith-based.

I do not mean to say that faith is never based on evidence. There may or may not be physical evidence to support the chosen belief, but if there is evidence, that evidence can be interpreted more than one way, and contrary evidence can be ignored or the interpretation of contrary evidence postponed. And that does not mean that all interpretations of the evidence are equally valid or that all possible choices are equally valid. I can be wise and logical in my choice to believe, or I can be foolish and illogical, and my choice could be wrong.

And if my belief is wrong, I may be convinced by logic or evidence that I am wrong, and I can change my belief. But it is always a choice. That is faith.

Some people will say that atheistic evolution is a "religion," but I would not go that far. To me, religion has to do with God.

But atheism, materialism, and evolution are *faiths*, chosen belief systems of those who believe them.

Faith can be evidence-based or not. Sometimes it is based on a personal relationship with a person you count as a trustworthy source of information.

My personal faith has three components, two based on evidence and one not. I believe, based on evidence, that God exists. I believe, based on evidence, that God inspired the Bible and the Bible is God speaking. But I *choose* to believe that God is a trustworthy source, that He will not lie to me, and therefore the Bible is true.

Believing *what God says* is a way of knowing by revelation—contrast with knowing by observation, which is the scientific method. I believe revelation from God is a *better* way of knowing than the scientific method of knowing from observation. God knows all things and will not lie. Therefore His word is true and sure, always, not like science and the scientific method, which is subject to mistakes.

Why do I say that science exercises faith when science itself does not hold on to the details of its teachings?

190 CREATION OF SPECIES

Science, in its inventory of discovered knowledge, is self-correcting (or should be in a healthy scientific environment). It can and does change its doctrines. Newton's laws of gravity were accepted for a long time but were eventually replaced by Einstein's general theory of relativity as fundamental law.

Science does not have faith in Newton's theory of gravity or Einstein's theories of relativity. But majority science has faith there is no God. Majority science has faith that evolution occurred.

But above all, science has faith in itself. It has faith that science and the scientific method are the best ways of knowing.

There is a faith in science. It is not faith in any particular theory. It is more fundamental than that. It is faith that science and the scientific method are the best ways of knowing things. Along with that, science and the methods of science are based on faith that consideration of supernatural causes, *even regarding origins and not repeatable processes,* is not necessary or helpful for determining truth. In other words, there is no God.

To say that consideration of supernatural causes of origins of the universe and life and the species of life is not necessary or helpful for discovering what is true is the same thing as saying there is no God, for if the possibility of God's existence is accepted, then consideration of supernatural causes certainly is necessary and helpful for determining what the truth is about origins.

The idea that science is the best way of knowing may seem self-evident to scientists and many non-scientists.

After all, isn't science the logical investigation of things based on observation and testing? And isn't it true that scientific knowledge becomes more accurate as time goes on, as new knowledge is discovered through experimentation and observation, as scientists correct their theories and knowledge in the face of evidence that shows the old knowledge needs to be changed? What is a better way of knowing than by observing and using reason to draw logical conclusions from what is observed?

Science using the scientific method is certainly a good way to know many things. Much of the technology we enjoy is based on knowledge science has discovered.

But there is an alternative. There is, what I believe, a better way of knowing.

A better way of knowing is to believe what God says in His word, the Bible.

Science has faith in itself, that it is the best way of knowing. I and many others have faith that God exists and that His word, the Bible, is true because God is trustworthy and will not lie to us.

Science believes in experimenting to find the truth. I believe in God's word, that it is true. I choose to believe what God says.

That is a great issue of controversy between Satan and God.

The Bible reveals that Lucifer was originally righteous as God created him. God must have instructed him in the right way to live. And as part of that instruction, God must have warned him of the consequences of sin, that sin would bring the penalties of suffering. We can know this because the Bible shows the mind of God, and God's way is to warn. The Bible is full of warnings from Genesis through Revelation. But Lucifer turned to sin (Ezekiel 28:15).

He chose the way of pride, vanity, conceit, and self-seeking. And as he sinned, he began to lose his wisdom, and he became more and more evil (Ezekiel 28:17).

Why did he do this? He has brought misery upon himself. He is awaiting final punishment, as the book of Revelation shows. God must have warned him what would happen if he went the way of vanity and sin. Yet, he chose that path anyway. Why? Did he want to be miserable?

One might say, he gave in to temptation. But there was no temptation as far as I can see in the Bible. Who would tempt him? If he was the first to sin, and he probably was, what evil being would tempt him into sin? And there was no evil tendency inside him to tempt him. God did not make him evil.

I don't think he wanted to be miserable. I don't think he foresaw all the suffering that has come upon him and will come upon him. Yet, God had warned him, so how could he not know the consequences?

The only explanation I can think of is that Lucifer simply did not believe God's warnings. He didn't trust that God was telling him the truth. He had no experience with sin and the suffering that comes from sin. He wasn't sure God was telling him the truth when God warned him that sin would bring suffering. During the time he lived righteously, he had known only happiness.

He had a mind that could think and reason, and he had free moral agency that could make decisions. He had great wisdom, and he understood what was at stake, at least in some sense (Ezekiel 28:12).

And if he didn't have faith in God's word and teaching, if he wasn't sure God was telling him the truth, he might have wondered if the way of vanity, pride, and self-seeking was a better way of life, a life that would be happier than the life he was living righteously.

This is speculation, but it fits with what God teaches in the Bible.

So Lucifer had a choice to make. He could believe God or he could try to find out for himself if vanity and sin would make him happier.

Maybe he thought to himself that he could ignore God's warnings and experiment with a little sin. He could try the way of vanity, pride, and self-seeking and see for himself if it brought him greater happiness.

So he performed the first scientific experiment. Instead of trusting and believing what God told him, he went the path of vanity, pride, and self-seeking. He sinned. But when he sinned, something happened to his mind. God's penalties came into effect. His mind—his wisdom—became corrupted. Sin may produce temporary pleasure, but it produces misery in the long term. Lucifer became more and more evil, more and more corrupt, and in the long term he did not find more happiness, but suffering came upon him as a penalty of sin.

Lucifer may have been the originator of the scientific method of experimentation, observation, and reasoning rather than believing God.

That is an issue today among human beings. Will we have faith in God, to believe His word, to trust Him that He will never lie to us but only tell us the truth, or will we have faith in the scientific method, the method of experimenting, observing results, and using our reason to draw conclusions, even if those conclusions contradict what God tells us?

God is teaching the human race lessons. He offers or will offer each one of us an opportunity to have eternal life in His kingdom as His son or daughter (2 Corinthians 6:18). But He wants children who will believe what He says, who will trust Him and have faith in His word, forever, and never doubt Him or "experiment" with sin as Lucifer did. He wants no more rebellions.

That is why faith is so important to God. That is why God counted Abraham's faith as righteousness. "Then He brought him outside and said, 'Look now toward heaven, and count the stars if you are able to number them.' And He said to him, 'So shall your descendants be.' And he believed in the LORD, and He accounted it to him for righteousness" (Genesis 15:5–6). See also Romans 4:3 and James 2:23.

Scientists claim they believe what they believe because of the scientific method, a method of research and experimentation, observation, and using reason to interpret the results and draw conclusions.

But that is *not* why they believe the scientific doctrines and knowledge they believe.

What has your science teacher directly observed? He or she claims that the knowledge they have comes from physical evidence, from observation. But it does not, in most cases.

How many fossils has your teacher of evolution personally dug up and directly observed? How many experiments has he or she done with DNA? More than ninety-nine percent of what your science teacher claims to know about science, including the theory of evolution, is based on the testimony of other scientists—not by the direct observation of your science teacher.

And they are asking you to believe on the same basis. They are not telling you to experiment and observe. They are telling you to trust and believe what other scientists have observed. They are telling you to trust and believe—have faith in—the words of your teacher and the words of your textbook.

But scientists who think this way are not consistent in the way they teach students. They advocate the scientific method of research, experimentation, observation, and interpretation, but they do not require science students to learn that way, but rather, they require students to believe what they are told. They do not require science students to learn by experimentation and observation, though there may be a few well-structured lab exercises to supplement classroom and textbook study. Almost everything the student learns in science classes is from the teacher's lectures and the textbooks.

In other words, students who believe their teachers when the teachers say, "evolution is a fact," are not practicing the scientific method of observation and interpretation. They are exercising faith in their teachers and their textbooks. Neither are they practicing experimentation and observation when they believe all the examples and explanations they read in their textbooks and hear from their teachers. All the explanations they read or hear are interpreted according to the biased thinking that evolution is true and there is no God. Evolution is the lens through which evidence is viewed and interpreted.

It is not a question of believing your own observations and experiments or believing what God says. It is a question of believing what atheistic scientists say or believing what God, the creator of the universe, says.

You are not practicing the scientific method when you swallow what is written in textbooks and spoken in classroom lectures.

Science teachers may claim that science is the best way of "knowing," but their students do not learn by the methods of science, which include experimentation, observation, testing, etc., but they learn by rote—by *believing* what their teachers and textbooks say. In other words, they take their teachers' word for things, that certain evidence has been observed and has been fairly and honestly interpreted. Yet they are ridiculed if they believe what God says more than what man says.

Who is more trustworthy? The God who created all things or the men who deny His existence? Whichever of the two you believe, you are believing *someone,* not necessarily observing for yourself according to the scientific method.

Actually, you are practicing the scientific method more by believing God because you are observing the universe directly yourself and drawing conclusions, that creation proves a Creator.

It is science and the scientific community that scientists have faith in. They have more faith and trust in themselves and each other than in God and His word.

But God himself is a witness to everything that ever happened. A scientist may dig up a fossil, document his observations, draw conclusions, and write it up in a textbook for you to read and trust and believe. But God was *there.* The scientist may observe a fossil after millions of years, but God was there and a witness when the animal or plant lived and died and became a fossil. God *knows* how the plants and animals came to exist. He knows everything. And He has documented the most important things to know about everything and given them to us in His word, the Bible.

There is nothing wrong with experimenting, observing, and learning from our observations, as long as this does not go against God's word. But God's word must come first.

Scientists trust other scientists and ask you to do the same, but God is looking for people who will trust Him first.

And sometimes you have to make a choice to trust and believe God or trust and believe what people say.

It is a matter of *believing*—having faith in—God and His word the Bible or the atheists who dominate the scientific community. If you disbelieve what God says, you cannot claim to believe the evidence, because you do not examine that evidence directly but only believe the materialistic scientists who filter and interpret the evidence. It is a matter of faith in God or faith in man.

Scientists can say, "We have examined the evidence and are eye-witnesses to what we have seen." But God is a greater witness because He *created* the evidence and He *knows* what it means.

So who will you believe, God or man?

That is your choice.

I choose to believe God first. God and His word, the Bible, is my chosen belief system—my faith. Science and my own observations and reasoning are ok, to a point, as long as they do not contradict God's word. But when God's word says something and science, or society, or my own reasoning tells me something else, I choose to trust and believe God. He is my trusted authority for what I believe.

I am convinced and believe that believing and trusting God's word is the best way of knowing things, not believing and trusting science and the scientific method of experimentation, observation, and human reasoning.

Science is *not* the best way of knowing things. Believing God is the best way of knowing things.

## Understanding the Bible

How can we understand the Bible?

Some people who have studied the Bible are concerned about what appears to them to be contradictions. But these are apparent contradictions, not real contradictions when properly understood. Right understanding often comes as a result of right attitude. The skeptic *looks* for contradictions and is glad when he thinks he has found them. He does not want God's word to make sense.

But someone with a right attitude toward God and His word looks for ways to reconcile different passages so they complement each other. And he often is able to do it.

But to correctly understand the Bible requires spiritual help from God. God designed it that way. He does not reveal His truth to everyone. God's Holy Spirit can help those whose minds are being opened by God to understand the Bible.

"No one can come to Me unless the Father who sent Me draws him; and I will raise him up at the last day" (John 6:44).

"But the Helper, the Holy Spirit, whom the Father will send in My name, He will teach you all things, and bring to your remembrance all things that I said to you" (John 14:26).

"For what man knows the things of a man except the spirit of the man which is in him? Even so no one knows the things of God except the Spirit of God. Now we have received, not the spirit of the world, but the Spirit who is from God, that we might know the things that have been freely given to us by God" (1 Corinthians 2:11–12).

The rest of the world cannot fully understand the Bible because their minds are blinded by Satan.

"So the great dragon was cast out, that serpent of old, called the Devil and Satan, who deceives the whole world; he was cast to the earth, and his angels were cast out with him" (Revelation 12:9).

God helps those who believe and obey Him to understand the truth.

"The fear of the LORD is the beginning of wisdom; A good understanding have all those who do His commandments" (Psalm 111:10).

If you are reading the Bible, and you find that God says something to you, and if you *believe* what God says and try your best to obey Him, then, when God sees your faith, He will help you understand more. And if you continue this process, believing and obeying God's word, God will open your mind to understand more and more of the Bible.

But if at some point you *disbelieve* what God tells you or fail to try to do what He says, the understanding can stop. God may no longer help you to understand His word, and your understanding will diminish.

Understanding God's truth is not a matter of college degrees or high IQ. It is a matter of believing and doing God's word. That is how God set it up.

Some are concerned about Bible dating vs. secular dating. They look at what historians say are various past dates, and see a difference between man-written history and the dates you can calculate from the Bible. Also, some are concerned about artifacts dated by radiocarbon (carbon-14) dating as older than described by Bible chronology.

But carbon-14 dating is dependent on a constant stream of cosmic ray radiation, one that has not varied for thousands of years. If that rate is not constant, carbon-14 dating is not accurate. If the rate of cosmic ray bombardment of the earth was less in the distant past, carbon-14 dates can appear too old, as I pointed out earlier in this book.

Tree ring dating can also be inaccurate, because in certain seasons, if there is a succession of dry and wet periods in a single year, trees can produce more than one ring per year, making them seem older.

As far as human history is concerned, there can be inaccuracies here too. Ancient nations and kings did not necessarily always keep their records honest, sometimes hiding unpleasant facts. Sequential dating can actually be overlapping dating, and what historians view as a sequence of periods of rule by different kings can include overlapping periods, making the historians' estimates too long.

In any case, it is for me a matter of believing man or God. I choose to believe God.

A key test of faith is, will we believe God *when we don't have all the answers?*

# Science vs. Technology

This society has developed a trust in science that is extreme, and wrong. Saying something is "scientific" is almost a synonym for saying it is true. If you say something is scientific, you are praising it, as people think.

Part of this admiration of science comes from inventions and technology. People are impressed with our modern push-button world of automobiles, airplanes, rockets to the moon and beyond, atomic power, electronics, medical advances and medicines, cell phones and computers, etc.

But what I have described is technology more than science. Science is a part of it. But technology is not entirely based on science. Science helps technology, but much of technology is based on trial and error. If a company is inventing a product, they may have hundreds of failures for each success in the development laboratory. Most of the public doesn't see that. They only see the final product and may imagine that everyone knew, based on science, how to do it right the first time. But that is not the case.

Product developers will try something, test it out, and if it doesn't work, try something else. If it works, they may try to make it work better. They don't perfectly know what they are doing. Scientific knowledge can help guide their efforts and give them ideas of what to try, but things don't always work as planned.

Some products never make it to market. They just don't work.

Don't be overly impressed with science because you see the results of trial and error in the products we enjoy. We only see the successes, not the far greater failures.

## God Is Creator

I know I talked about this already, but I want to repeat it for emphasis.

I am sure that what I am saying will be misrepresented by some who disagree. They will accuse me of denying God's role as creator by saying, angels, not God, designed and created the species. That is emphatically *not* what I am saying. I am saying that God is creator, *God* designed the species, but He did it *through* the angels, using angels as His servants to work out the details under God's authority. But God is still the creator, though He uses His servants as tools in His hand. This cannot be overemphasized.

Notice this passage again. "Therefore, when the Lord knew that the Pharisees had heard that Jesus made and baptized more disciples than John (though Jesus Himself did not baptize, but His disciples), He left Judea and departed again to Galilee" (John 4:1–3).

Christ builds His church (Matthew 16:18), but He uses the human instruments of the disciples to help do it

(John 21:15–17; 1 Corinthians 3:9–10; Ephesians 4:11–16). Also, each of the members, as a part of the church, has a part to play (1 Corinthians 12:27). Notice also, that the church Christ builds is *not perfect,* and part of the reason is that Christ uses imperfect instruments to build it.

Moses judged Israel, but he used an organization of judges to help him do it (Exodus 18:13–26). Solomon built the temple (1 Chronicles 17:11–13; 1 Kings 6:1–2), but He did not do the work directly with his own two hands—he used servants and workmen to do it (1 Kings 5:2–18). Christ will rule His kingdom on the earth, but the saints will rule with Him (Revelation 2:26–27; 3:21; 20:4–6; Luke 19:11–19; Matthew 19:27–28; 25:14–21).

This is the principle of delegation, and it is described and used many places in the Bible, and it is used in our time in business, government, and the military.

God designed and created the species, but He *delegated* part of that work to the angels. Being limited, they took shortcuts, and being imperfect after they sinned, their work was imperfect (just as the church Christ builds is imperfect because He uses imperfect human instruments and imperfect human materials), but God was able to use most of the work they produced for His purposes.

# Relationship between GMB and Intelligent Design

What I call "Guided Modification and Branching" (GMB) is a religious flip side of intelligent design. It is not intelligent design, for intelligent design does not examine the nature of the intelligence that did the designing. ID only examines the evidence that intelligent design did occur. Intelligent design is not religious.

You can think of GMB as a link between ID and the Bible, but it is not ID itself.

What I am proposing is an explanation from a religious and biblical point of view of why and how intelligent design occurred as it did—and who did the designing.

It was God who did the designing, but in part through angels, who acted as His servants—His agents—and tools in His hand doing some of the designing. And because the angels are limited and imperfect, the design of species does not show the kind of perfection we might expect if God did everything Himself, directly, in an instant.

## Summary

The theory of evolution can explain much of the evidence better than traditional creationism that says that the earth is only six thousand years old, but evolution does not explain everything well. The irreducible complexity that exists in nature is not easily explained by the theory of evolution, but intelligent design explains it perfectly.

The theory of evolution cannot be proved, yet the secular scientific community pushes acceptance of it on society. Not only is evolution unproved, there is evidence that it could not have occurred. That evidence includes irreducible complexity and missing transitional forms in the fossil record.

There is evidence for the emergence of species through common descent over millions of years. That evidence is primarily the evidence of the fossil record, vestigial organs, and DNA, particularly non-functional DNA.

Intelligent design to modify DNA to produce new species from parent species, combined with emergence of species through common descent over millions of years, can explain all the physical evidence science has found.

The truth is, for those who are willing to look at the evidence objectively, there is evidence of intelligent design, but *limited, imperfect design,* not sudden, perfect design done directly by an infinite God.

But science refuses to consider intelligent design because its culture and scientific method, as practiced, does not allow consideration of supernatural causes, which is what atheistic science says intelligent design implies. Science blindly rejects God and His influence in the natural world.

But students and all people who study science can make up their own minds about this apart from what the scientific community wants everyone to believe.

For those who believe in God, a question arises. If God intelligently designed the species, why would He use such a slow method of common descent of the species over millions of years? With His infinite mind and power, He could design and create all the species at once. Also, why would He create a cruel and competitive ecology when the Bible shows that He desires peace and harmony in the natural world?

This can be explained by realizing that God, though infinite and perfect in wisdom, power, and righteousness Himself, sometimes works through imperfect agents—servants—as He does in the church and in the world and as documented in the Bible. He shares His work and accomplishments with His servants out of love, to give them work to do, to develop and test their character, and to give them a sense of accomplishment and joy in using their abilities in God's service.

I suggest that God used some of the angels, specifically Lucifer and that portion of God's angels He placed on the earth, to make the changes to the DNA of parent species to make new species. The angels, though powerful, are not infinite like God, and it would take them time to make the changes needed and to manage a growing ecology.

God has used imperfect angelic instruments—angels, many of whom sinned and became demons—to develop the design of the species to this day. Angels are limited.

Their minds and their intelligence are not infinite like God's mind. They cannot do everything at once. They can change life forms a little bit at a time over millions of years. They may take shortcuts, making the easiest possible change to get a result. And after they sinned, the species they developed were no longer according to a righteous plan that God originally intended. Instead of cooperation and harmony and peace in the animal kingdom, as God describes as the ideal in the Bible, there is cruelty in the animal kingdom with animals killing animals in competition or for food.

Lucifer and the angels on the earth started out on the right track before Lucifer sinned and led many of the other angels to sin. But Lucifer and the angels had free moral agency, and they sinned and rebelled against God. Lucifer became Satan, God's enemy, and his angels became demons. But God did not remove Lucifer from his office, and the work of developing species continued.

A savage and competitive ecology was developed, with predator animals killing other animals, which was not God's original intent at the time. Huge dinosaurs were developed that would make it difficult or impossible for man to live on the earth. God allowed this to continue for a time because it suited His purpose, occasionally intervening, such as bringing about the extinction of the dinosaurs. But eventually the time came for this to end, and the surface of the earth was destroyed.

Then in six days, as related in the first chapter of Genesis in the Bible, God renewed the surface of the earth, preparing it for life and for man, making the same plants and animals that had existed before the destruction according to the same design and DNA code. He did not waste the design of life embedded in the DNA code but used it for His purpose. And in the sixth day, God made man, a new creature, making him similar to man-like creatures that existed before, but with a human spirit in him that enhanced the power of his mind, making him vastly superior to the animals and allowing him to have a relationship with God that animals cannot have.

God has allowed a competitive, savage natural world at this time for the same reason He allows suffering and injustice in man's world. Why? He is teaching lessons. He wants the human race to get a bellyful of Satan's wrong way of life. He wants mankind to see the horrific end result of this evil system and way of life, which is yet to come in the near future. He wants us to experience it firsthand in a way we will never forget, individually or as a race of beings.

But then, God will teach mankind the right way of life, and we will experience that too. We will see that God's way of love and peace brings happiness. And we will see the contrast and learn the lesson—for eternity. Then, none of us will imagine, not in millions of years, that Satan's way of vanity and hatred and violence is better.

All this is consistent with the whole Bible and a literal understanding of Genesis and with all the physical evidence science has found.

I cannot prove that the species and kinds of life developed this way, but it is consistent with physical and

Bible evidence, and it is the only explanation I know of that is consistent.

Some sincere people have difficulty accepting the Bible because they have not known how to reconcile it with scientific evidence.

It is my hope that the ideas I present in this book may cause those who have rejected the Bible for that reason to reconsider. There is evidence that God exists and the Bible is His word, and the Bible does not contradict any true scientific evidence.

## The Future Prospects of GMB

Will I convince anyone with this book? Will my suggestions, or anything in the intelligent design movement for that matter, catch on?

We can only wait and see.

But the older generation and the establishment is not likely to accept what I am saying. Some will. Those in the establishment who have careers in science or religion may consider what I say without bias, those who have an open mind, and I am sure some will.

But those with the best chance of honestly considering what I say in this book are the youth and those who are still questioning, still looking for answers, still investigating the big questions of life—those who are not already entrenched in a worldview they are not willing to change.

I think it was Max Planck who said that new scientific theories do not replace old ones because establishment scientists give up their old theories to accept new ones. He said that new theories catch on when the older scientists die out and new, younger scientists accept the new theories.

Perhaps something similar will happen here.

New ideas are not always introduced in scientific papers published in mainstream scientific journals. Whatever monopoly or near-monopoly those journals have had on scientific discussion and thought is rapidly eroding in the age of the Internet and the multitude of books being published and discussed. The Internet, public lectures, and books are becoming the new forum for new ideas.

I know of no other way but what I have suggested in this book to reconcile the Bible with physical evidence science has discovered. In fact, even apart from the Bible, I know of no other way to interpret the physical evidence science has discovered logically and honestly. Genetics, chemistry, anatomy, and the fossil record strongly suggest common descent. But transitional gaps and the improbability of life and major species emerging solely through random mutation and natural selection point to intelligent design. And the imperfections in the natural world point to God using imperfect servants to do the work of designing under His authority and supervision.

The scientific evidence points to intelligent design by limited, imperfect designers, and the Bible allows for this and even supports it.

If anyone has a better explanation, I am open to consider it.

My proposal in this book, which I call Guided Modification and Branching (GMB), will probably not reconcile most atheists and militant promoters of the theory of evolution to the Bible, nor will this make peace between the two sides. But it can help our youth and those who are impartial to understand that the Bible does not contradict physical evidence science has discovered.

Majority science, dominated by militant atheists intent on imposing their atheistic faith on society, will continue to war against any belief in God, the supernatural, or intelligent design.

Because the science community has allowed itself to be dominated by militant atheism, it may be losing credibility with the public. People see the bias against God that exists in science. They see the persecution of scientists who do not accept evolution. And they must wonder, if the science community had the evidence on their side, why the persecution to marginalize those who have a different opinion? If the evidence is there, unfair persecution in an attempt to silence critics would not be necessary. Thus, persecution and obvious bias have the effect of working against the theory of evolution by damaging the credibility of evolutionists.

The real conflict is theological, even with evolutionists. The science of biology and the theory of evolution are peripheral. Atheists use science as a tool to advance their faith in atheism (and impose it on others). Science isn't the main issue—religion is. Evolutionists and atheists are more influenced than they admit by the arguments that God cannot exist because He would not allow evil (or imperfect design of life) to exist.

How big is the issue of imperfect or cruel design in the evolution vs. creation controversy? It is much bigger than most people on either side of the argument realize. It is *huge*.

It is a big factor in atheism and naturalism in general and evolution in particular. It, along with the suffering and injustice of this world, may be a big factor in why atheists are atheists.

It is a big factor in how and why atheists have been able to take virtual control over the scientific community, the education system, and the traditions and rules of science to cause science to reject as myth any possibility of God's intervention in the natural world.

It is a bigger factor than you would think from reading scientific papers and textbooks. In official writings, science must exclude supernatural considerations even to the extent of not discussing *why* they believe *there is no supernatural.*

But it is in the mind of atheistic evolutionists and naturalists, and it comes out in debates and books for the general public.

In scientific papers and public school textbooks, evolutionists must not mention God or the supernatural.

They must be consistent in their position that to mention even the possibility of supernatural intervention is outside the rules of science and would be bringing religion into science and the science classroom. They must be consistent in this, lest they leave an opening for intelligent design proponents to bring ID into the classroom. So you won't read in a secular textbook or paper published in a science journal that God could not have designed the species because species are imperfectly designed and nature is cruel, contrary to the idea of a perfect, infinite Creator.

But in popular books, where the *real* debate takes place, the gloves are off, for both sides.

Just as ID scientists can publish books teaching intelligent design, so atheistic evolutionists can speak their minds on the theological side and say that God could not have created life and the species of life because the imperfection and cruelty of life are incompatible with a perfect and infinite creator.

I think even Darwin was concerned about this.

But God is working out a purpose, and that purpose includes teaching the human race lessons about the right way of life, and it suits that purpose to allow an unjust, evil world at this time. But that will soon change.

God exists. You can prove that the Bible is His inspired word. He has intervened in physical processes, first to design and make this universe and then to develop the species of life on this earth, probably using angels as His intelligent servants to help in the design of life.

Every human will eventually have to make a choice to believe the word of the Creator of the universe, or not.

That is your choice as well.

# Selected Bibliography

Behe, Michael J. *Darwin's Black Box: The Biochemical Challenge to Evolution*. New York: Simon & Schuster, Inc., 2006.

Berra, Tim M. *Evolution and the Myth of Creationism: A Basic Guide to the Facts on the Evolution Debate*. Stanford, CA: Stanford University Press, 1990.

Black, Jim Nelson. *When Nations Die: America on the Brink: Ten Warning Signs of a Culture in Crisis*. Wheaton, IL: Tyndale House Publishers, Inc., 1994.

Blackmore, Susan. *Consciousness: An Introduction*. New York: Oxford University Press, 2004.

Brockman, John, ed. *Intelligent Thought: Science vs. the Intelligent Design Movement*. New York: Vintage Books, 2006.

Carter, Robert, ed. *Evolution's Achilles' Heels*. Powder Springs, GA: Creation Book Publishers, 2015.

Chalmers, David J. *The Conscious Mind: In Search of a Fundamental Theory*. Oxford: Oxford University Press, 1996.

Collins, Francis S. *The Language of God: A Scientist Presents Evidence for Belief*. New York: Simon & Schuster, Inc., 2006.

Cowles, Henry M. *The Scientific Method: An Evolution of Thinking from Darwin to Dewey*. Cambridge: Harvard University Press, 2020.

Coyne, Jerry A. *Why Evolution Is True*. New York: Penguin Group, 2009.

Crick, Francis. *The Astonishing Hypothesis: The Scientific Search for the Soul.* New York: Simon & Schuster Inc., 1994.

Custance, Arthur C. *Without Form and Void: A Study of the Meaning of Genesis 1:2.* Windber, PA: Classic Reprint Press, 1970.

Dankenbring, William F. *The First Genesis: A New Case for Creation.* Altadena, CA: Triumph Publishing Company, 1975.

Dawkins, Richard. *The Blind Watchmaker: Why the Evidence of Evolution Reveals a Universe Without Design.* New York: W. W. Norton & Company, Inc., 1986.

Dembski, William A. *Intelligent Design: The Bridge Between Science & Theology.* Downers Grove, IL: InterVarsity Press, 1999.

Dembski, William A., ed. *Mere Creation: Science, Faith & Intelligent Design.* Downers Grove, IL: InterVarsity Press, 1998.

Dembski, William A. *No Free Lunch: Why Specified Complexity Cannot Be Purchased without Intelligence.* Lanham, MD: Rowman & Littlefield Publishers, Inc., 2002.

Dembski, William A. Jonathan Wells, *The Design of Life: Discovering Signs of Intelligence in Biological Systems.* Dallas: The Foundation for Thought and Ethics, 2008.

Dembski, William A. *The Design Revolution: Answering the Toughest Questions about Intelligent Design.* Downers Grove, IL: InterVarsity Press, 2004.

Dembski, William A. *The End of Christianity: Finding a Good God in an Evil World.* Nashville: B & H Publishing Group, 2009.

Dembski, William A., ed. *Uncommon Dissent: Intellectuals Who Find Darwin Unconvincing*. Wilmington, DE: ISI Books, 2004.

Dembski, William A. and Sean McDowell. *Understanding Intelligent Design: Everything You Need to Know in Plain Language*. Eugene: Harvest House Publishers, 2008.

Dennett, Daniel C. *Consciousness Explained*. New York: Little, Brown and Company, 1991.

Denton, Michael. *Evolution: A Theory in Crisis*. Chevy Chase, MD: Adler & Adler, Publishers, Inc., 1985.

Dykstra, Pieter. *Creation and the Consequence of Satan's Fall: An exposition of the controversial "Gap Theory" as found in Genesis 1:1 and Genesis 1:2*. Bloomington, IN: WestBow Press, 2012.

Eldredge, Niles. *Reinventing Darwin: The Great Debate at the High Table of Evolutionary Theory*. New York: John Wiley & Sons, Inc., 1995.

Feyerabend, Paul. *Against Method*. Brooklyn: New Left Books, 2010.

Feyerabend, Paul. *Science in a Free Society*. London: Verso, 1978.

Fields, Weston. *Unformed and Unfilled: A Critique of the Gap Theory*. Green Forest, AR: Master Books, Inc., 2005.

Flew, Antony. *There Is a God: How the World's Most Notorious Atheist Changed His Mind*. New York: HarperCollins Publishers, 2007.

Futuyma, Douglas J. *Science on Trial: The Case for Evolution*. Sunderland, MA: Sinauer Associates, Inc., 1995.

Giberson, Karl W. *Saving Darwin: How to Be a Christian and Believe in Evolution*. New York: HarperCollins Publishers, 2008.

Gonzalez, Guillerno and Jay W. Richards. *The Privileged Planet: How Our Place in the Cosmos Is Designed for Discovery*. Washington, DC: Regnery Publishing, Inc., 2004.

Hayward, Alan. *Creation and Evolution: Rethinking the Evidence from Science and the Bible*. Minneapolis: Bethany House Publishers, 1985.

Heeren, Fred. *Show Me God: What the Message from Space Is Telling Us About God*. Wheeling, IL: Searchlight Publications, 1995.

Hitching, Francis. *The Neck of the Giraffe: Where Darwin Went Wrong*. New Haven, CT: Ticknor & Fields, 1982.

Hossenfelder, Sabine. *Lost in Math: How Beauty Leads Physics Astray*. New York: Basic Books, 2018.

Jastrow, Robert. *God and the Astronomers*. New York: W. W. Norton & Company, Inc., 1992.

Johnson, Phillip E. *Darwin on Trial*. Downers Grove, IL: InterVarsity Press, 1993.

Johnson, Phillip E. *Defeating Darwinism by Opening Minds*. Downers Grove, IL: InterVarsity Press, 1997.

Johnson, Phillip E. *Objections Sustained: Subversive Essays on Evolution, Law & Culture*. Downers Grove, IL: InterVarsity Press, 1998.

Johnson, Phillip E. *Reason in the Balance: The Case Against Naturalism in Science, Law & Education*. Downers Grove, IL: InterVarsity Press, 1995.

Koch, Christof. *The Quest for Consciousness: A Neurobiological Approach*. Englewood, CO: Roberts and Company Publishers, 2004.

Krauss, Lawrence M. *A Universe from Nothing: Why There Is Something Rather than Nothing*. New York: Simon & Schuster, Inc., 2012.

Kupelian, David. *The Marketing of Evil: How Radicals, Elitists, and Pseudo-Experts Sell Us Corruption Disguised as Freedom*. Washington DC: WND Books, 2015.

Langford, Jack W. *The Gap Is Not a Theory: An Examination of the First Chapter of Genesis*. 2011.

Lewin, Roger. *Bones of Contention: Controversies in the Search for Human Origins*. New York: Simon & Schuster, 1987.

Margenau, Henry and Roy Abraham Varghese, eds. *Cosmos, Bios, Theos: Scientists Reflect on Science, God, and the Origins of the Universe, Life, and Homo sapiens*. Peru, IL: Open Court Publishing Company, 1997.

Mayr, Ernst. *What Evolution Is*. New York: Basic Books, 2001.

McGrath, Alister. *Dawkins' God: Genes, Memes, and the Meaning of Life*. Malden, MA: Blackwell Publishing, 2007.

Meyer, Stephen C. *Darwin's Doubt: The Explosive Origin of Animal Life and the Case for Intelligent Design*. New York: HarperCollins, 2013.

Meyer, Stephen C. *Signature in the Cell: DNA and the Evidence for Intelligent Design.* New York: HarperCollins, 2009.

Milton, Richard. *Shattering the Myths of Darwinism.* Rochester, VT: Park Street Press, 1997.

Nelson, Paul, Robert C. Newman, and Howard J. Van Till. *Three Views on Creation and Evolution.* Grand Rapids: Zondervan, 1999.

Penrose, Roger. *Fashion, Faith, and Fantasy in the New Physics of the Universe.* Princeton, NJ: Princeton University Press, 2016.

Perloff, James. *The Case Against Darwin: Why the Evidence Should Be Examined.* Burlington, MA: Refuge Books, 2002.

Perloff, James. *Tornado in a Junkyard: The Relentless Myth of Darwinism.* Burlington, MA: Refuge Books, 1999.

Powers, Kirsten. *The Silencing: How the Left Is Killing Free Speech.* Washington, DC: Regnery Publishing, 2015.

Ross, Hugh. *Creation and Time: A Biblical and Scientific Perspective on the Creation-Date Controversy.* Colorado Springs, CO: NavPress Publishing Group, 1994.

Ross, Hugh. *The Creator and the Cosmos: How the Greatest Scientific Discoveries of the Century Reveal God.* Colorado Springs, CO: NavPress, 2001.

Sailhamer, John H. *Genesis Unbound: A Provocative New Look at the Creation Account.* Colorado Springs, CO: Dawson Media, 2011.

Scott, Eugene C. *Evolution vs. Creationism: An Introduction.* Berkeley and Los Angeles: University of California Press, 2009.

Searle, John R., including exchanges with Daniel C. Dennett and David J. Chalmers. *The Mystery of Consciousness.* New York: The New York Review of Books, 1997.

Simmons, Geoffrey. *What Darwin Didn't Know.* Eugene, OR: Harvest House Publishers, 2004.

Smolin, Lee. *The Trouble with Physics: The Rise of String Theory, the Fall of a Science, and What Comes Next.* New York: Houghton Mifflin Company, 2006.

Snoke, David. *A Biblical Case for an Old Earth.* Grand Rapids: Baker Books, 2006.

Spetner, Lee M. *Not by Chance: Shattering the Modern Theory of Evolution.* Brooklyn: The Judaica Press, Inc., 1997.

Strobel, Lee. *The Case for a Creator: A Journalist Investigates Scientific Evidence That Points toward God.* Grand Rapids: Zondervan, 2004.

Susskind, Leonard. *The Cosmic Landscape: String Theory and the Illusion of Intelligent Design.* New York: Back Bay Books, 2006.

Ward, Peter D. and Donald Brownlee. *Rare Earth: Why Complex Life Is Uncommon in the Universe.* New York: Copernicus Books, 2000.

Wells, Jonathan. *Icons of Evolution: Science or Myth? Why Much of What We Teach About Evolution is Wrong.* Washington DC: Regnery Publishing, Inc., 2000.

Wells, Jonathan. *The Politically Incorrect Guide to Darwinism and Intelligent Design*. Washington DC: Regnery Publishing, Inc., 2006.

Whitcomb, John C. and Henry M. Morris. *The Genesis Flood: The Biblical Record and Its Scientific Implications*. Phillipsburg, NJ: P&R Publishing Company, 2011.

Woodmorappe, John. *Noah's Ark: A Feasibility Study*. Dallas: Institute for Creation Research, 1996.

Woodward, Thomas. *Darwin Strikes Back: Defending the Science of Intelligent Design*. Grand Rapids: Baker Books, 2006.

Made in the USA
Monee, IL
26 April 2022

95423775R00128